A CONCEPT OF AGRIBUSINESS

John H. Davis

Director, Program in Agriculture and Business

and

Ray A. Goldberg

Assistant Professor of Business Administration

Martino Fine Books
Eastford, CT
2021

Martino Fine Books
P.O. Box 913,
Eastford, CT 06242 USA

ISBN 978-1-68422-524-8

Cover Design Tiziana Matarazzo

Printed in the United States of America On 100% Acid-Free Paper

A CONCEPT OF AGRIBUSINESS

John H. Davis

Director, Program in Agriculture and Business

and

Ray A. Goldberg

Assistant Professor of Business Administration

DIVISION OF RESEARCH

GRADUATE SCHOOL OF BUSINESS ADMINISTRATION

HARVARD UNIVERSITY · BOSTON · 1957

Library of Congress Catalog Card No. 57-8929

Printed at
The Alpine Press, Inc.
Boston, Massachusetts, U.S.A.

*"The agricultural world and the industrial world are not two separate economies having merely a buyer-seller relationship. Rather, they are so intertwined and inseparably bound together that one must think of them jointly if there is to be any sound thinking about either one or the other."**

*** Address by T. V. Houser, Chairman of the Board, Sears, Roebuck and Co., at the National Institute of Animal Agriculture Conference, April 15, 1955.**

ADVISORY COMMITTEE TO PROGRAM IN AGRICULTURE AND BUSINESS

CURRENT COMMITTEE

CHAIRMAN:
Charles R. Sayre
President, Delta and Pine Land Company

VICE CHAIRMAN:
James A. Moffett, 2nd
Vice President
Corn Products Refining Company

VICE CHAIRMAN:
Frank J. Welch
Dean, College of Agriculture and Home Economics, University of Kentucky

---

Carl H. Black
Former Chairman of the Board, American Can Company

William Rhea Blake
Executive Vice President, National Cotton Council of America

Donald K. David
Dean Emeritus, Harvard Business School

Homer R. Davison
Vice President, American Meat Institute

Charles Wesley Dunn
President, The Food Law Institute

William J. Fisher
Vice President, The Oliver Corporation

Clarence Francis
Director, General Foods Corporation

Rudolph K. Froker
Dean, College of Agriculture, University of Wisconsin

J. Kenneth Galbraith
Professor of Economics, Harvard University

Raymond W. Miller
Consultant on Rural Affairs
Lecturer, Harvard Business School

Edwin B. Mitchell
Lobeco, South Carolina

William I. Myers
Dean, New York State College of Agriculture, Cornell University

Lansing P. Shield
President, The Grand Union Company

C. Leigh Stevens
President, C. L. Stevens Company
Lecturer, Harvard Business School

Don A. Stevens
Vice President, General Mills, Inc.

Jesse W. Tapp
Chairman of the Board, Bank of America

Kenneth W. Taylor
Deputy Minister, Department of Finance, Dominion of Canada

Arthur O. Wellman
President, Nichols and Company, Inc.

Oris V. Wells
Administrator, Agricultural Marketing Service, U. S. Department of Agriculture

FORMER MEMBERS

James F. Bell
Chairman of the Committee on Finance and Technological Progress, General Mills, Inc.

Cason J. Callaway
Former President and Chairman of the Board, Callaway Mills

Chester C. Davis
Regents Professor, University of California, Berkeley

John H. Davis
Former Assistant Secretary of Agriculture

R. G. Gustavson
Former Chancellor, University of Nebraska

John Holmes
Chairman of the Board, Swift and Company

A. King McCord
President, The Oliver Corporation

Morris Sayre (deceased)
Vice Chairman of the Board, Corn Products Refining Company

Foreword

THIS VOLUME represents the first publication growing out of research conducted under the Program in Agriculture and Business which was formally undertaken at the Harvard Business School in December 1952. The underlying concept of the present program first took concrete form in 1944 when the President and Fellows of Harvard College authorized the establishment of The Food Foundation under the administration of the Business School. Donald K. David, then Dean of the School, in his years as president of a company processing agricultural products had developed a strong conviction as to the need for better understanding of the mutually supporting relationships between agriculture and business and a belief that research and teaching at the School could contribute significantly to this understanding. The purposes of the Foundation were (1) to carry on fundamental research relating to the production, distribution, and use of agricultural and processed food products; (2) to develop fundamental economic policies governing the production, distribution, and use of such products; (3) to study and, as possible, to improve fundamental economic relationships among all engaged in the production, distribution, and use of such products; and (4) thus and otherwise to improve the food economy of the American people.

The Food Foundation itself did not reach fruition, in part because of difficulties in financing its program on a suitable scale and in part because of the pressure of other matters at the School in the immediate postwar period. The desire of the School to make teaching and research in the area of agriculture and business relationships a part of its over-all program continued, however, with undiminished strength. In 1950 the basic idea gained new impetus when a generous gift of the late George M. Moffett provided endowment for a Professorship in Agriculture and Business. Again, however, the launching of the program was delayed because of the diversion of a substantial fraction of our personnel to research on problems facing government agencies during the Korean conflict.

In the fall of 1952 development work for the program moved ahead rapidly, and Dean David invited a number of distinguished leaders from business, agriculture, and academic life who were interested in the problems of agricultural-business relationships to serve as an advisory committee to the School in regard to this program. In December 1952 this advisory committee met at the School to review the objectives which had been developed for the program and the plans to implement such objectives. The broad purposes of the program as finally adopted by the committee and by the School were stated as being to "conduct studies of agricultural and industrial relationships through analyses of technical, economic, and human factors which govern these relationships, particularly the decision-making points, and stimulate sound action in the light of these studies so that industry and agriculture may contribute most efficiently toward meeting their responsibilities in our growing economy."

At the time the program was launched, Dean David, who had worked diligently over many years to establish this program at the School, stated that "there are relatively few manufacturing businesses in the nation today which do not rely on agriculture for a portion of their own materials. The number of nonmanufacturing enterprises closely associated with, or dependent upon, agriculture is also impressive. The extent of this interdependence is commonly overlooked. The impact of management decisions on the farm operation has all too seldom been a cause for study or analysis. It is hoped that by undertaking studies in the basic technical, economic, and human aspects of the relationships between these two primary segments of our economy, it will be possible to influence a greater degree of coordination, to the end that agriculture and its related supplying, processing, and distributing industries will more effectively serve the needs of a fast-growing population."

The first major study undertaken under this program had as its objective a description in quantitative terms of the character and extent of the existing interrelationships between agriculture and the industries which supply agriculture and which process and distribute the products of agriculture. There is a two-way interdependence between businessmen and farmers in the dual roles of suppliers and purchasers. The interdependence is so close that the authors have coined a new word, "Agribusiness," to describe the interrelated functions of agriculture and business. This volume sets forth the structure of agribusiness in the United States and the nature and magnitude of the relationships that exist among the many interrelated parts of agriculture and business. Thus the study provides a frame of reference for those in studying domestic food and fiber operations and suggests a type of approach that may prove fruitful in studying problems involved in these areas.

This program owes a great deal to the contributions of the members of the advisory committee which Dean David set up at the time the program was inaugurated and especially its chairman, Mr. Charles R. Sayre. The School is indebted to each of them.

Financial support for this study has come from a number of sources. The original endowment by Mr. George M. Moffett for a Professorship at the School in Agriculture and Business has been used in part for this research. In addition, a major gift in support of the program in Agriculture and Business was made by the Whitehall Foundation. Other gifts in support of the program have come from the Southern States Cooperative, Inc., the Cooperative Grange League-Federation Exchange, Inc., the Farm Bureau Cooperative Association, Inc. (Ohio), and the General Foods Corporation. The Business School is truly grateful for this financial support which made this study possible, and which will lead to further studies of the relationships of agriculture and business.

STANLEY F. TEELE
Dean

BERTRAND FOX
Director of Research

Soldiers Field
Boston, Massachusetts
February 1957

Acknowledgments

THIS IS the first study to be published as the result of research under the recently established Program in Agriculture and Business at the Graduate School of Business Administration, Harvard University. Within the Business School this project serves as a basis for orienting the teaching, research, and extension phases of the Program. It is the hope of the authors that other schools of business, colleges of agriculture, farm leaders, businessmen, government officials, and others will find the study useful and challenging.

The authors are indebted to numerous persons for their assistance and cooperation. Among their associates at the Business School, special acknowledgment is due Donald Kirk David who, while Dean of the Business School, provided the inspiration for starting the Program in Agriculture and Business; Dean Stanley F. Teele whose keen interest and counsel have helped to guide the study; and Professor Bertrand Fox, Director of Research, whose resourcefulness, constructive advice, and suggestions have been invaluable. Special gratitude is due George S. Peer who, as Executive Secretary of the Program in Agriculture and Business, handled many of the administrative problems and details connected with the study; and William Applebaum, Visiting Consultant on Food Distribution, who offered many helpful suggestions. The authors also are appreciative for the sound advice of Dr. Charles R. Sayre and other members of the Advisory Committee to the Program in Agriculture and Business.

The authors are indebted to Professor Robert L. Clodius of the University of Wisconsin and Herman M. Southworth of the Agricultural Marketing Service, United States Department of Agriculture, both of whom took leave from their regular work to assist in assembling some of the basic material, to Professor Wassily W. Leontief, Professor James M. Henderson, Dr. Elizabeth W. Gilboy, and their associates in the Harvard Economic Research Project whose generous counsel and cooperation did much to facilitate the input-output analysis presented in Chapter 3; and to O. V. Wells, Administrator of the Agricultural Marketing Service, United States Department of Agriculture, and his staff for their counsel and assistance in assembling data.

The authors wish to thank the following individuals for the help given through various stages of this study: Miss Ruth Norton, Secretary of the Division of Research, for her expert assistance in guiding this study from manuscript to publication; Miss Ruth E. Davison for her invaluable secretarial assistance; Mrs. Rose Kneznek and Miss Caroline Timmerman of the Bureau of Business Research for their meticulous and accurate presentation of the tables and charts; Miss Anabelle Forsch and Mr. John Head for the artwork; and Miss Yvette St. Jean for the design of the book.

Finally, the authors wish to thank their wives, Edna and Thelma, for all their help, inspiration, and patience.

While all the above people and many others share in making this publication possible, the authors take the full responsibility for the data and opinions reported herein.

JOHN H. DAVIS
RAY A. GOLDBERG

Soldiers Field
Boston, Massachusetts
February 1957

Table of Contents

List of Exhibits

List of Text Tables

List of Appendix Tables

Introduction

The Setting

Today our food and fiber economy is being reshaped by the expanding forces of science and technology. Not only are the functions of farming, processing, and distribution all undergoing great physical evolution, but unprecedented change is taking place throughout our rural society—socially, politically, morally, and even spiritually. Virtually nothing is being left untouched. Out of this technological revolution have arisen innumerable problems which as yet are unsolved. These problems are further complicated by the fact that food and fiber operations are a part of a total national economy which itself is undergoing change at a rapid rate.

Fundamentally, today's food and fiber problems—particularly on the farm—are rooted in imbalances brought about by spotty progress on an uneven front. Certain phases of this segment of the economy have surged ahead rapidly, others at a more moderate rate, and some at a snail's pace. Illustrative of this irregular type of progress are the following examples:

Plant and animal production have been expanded almost 50% in two decades without the development of corresponding market outlets.

Agriculture has been confronted with firmer costs of production, transportation, processing, and distribution without the development of corresponding behavior in the price structure of farm products.

The need has been created for larger farm units, but some three million farm families remain on units which are too small for efficient operation.

The need for increased skills, technical know-how, and managerial ability on the part of the farm operator has been created without developing adequate on-the-job training programs or other means of enabling such operators to keep abreast of the times.

The interdependence of agriculture and business has been increased without creating adequate machinery whereby these factors of the economy can plan and work together in formulating sound policies which are mutually beneficial to them and which further our national economic goals.

The result of this uneven progress is to throw our rural economy badly out of balance, both internally and in relation to the national economy. This imbalance appears in the form of hundreds of problems, large and small. These not only complicate the lives of rural families but are expensive to the public, both in terms of direct costs and in terms of lowering our national economic potential. When viewed in broad perspective, these problems are seen to converge about two poles: (1) that of commercial agriculture, and (2) that of low-income farmers. Each is a gigantic complex of problems related to the other, yet basically their respective solutions lead in somewhat divergent directions—the former falling chiefly in the field of business economics, and the latter in that of economic growth and development.

Agribusiness

The concept of agriculture as an industry in and of itself or as a distinct phase of our economy was appropriate 150 years ago when the typical farm family not only raised crops and livestock but also produced its own draft animals, tools, equipment, fertilizers, and other production items; processed its own food and fiber; and retailed in the community most of the excess above family needs. Then virtually all operations relating to growing, processing, storing, and merchandising food and fiber were a function of the farm. This being the case, it was appropriate to think of all such things as within the scope of the meaning of the word "agriculture."

As is explained more fully in Chapter 1, along with the technological revolution has come a narrowing of the function of the farm. Basically, farming has changed from a subsistence to a commercial status in that now the progressive farm family consumes only a fraction of what it grows—the balance being sold to feed the 88% of the population employed off the farm. The modern farmer is a specialist who largely confines his operations to growing crops and livestock. The functions of storing, processing, and distributing food and fiber have been transferred in large measure to off-the-farm business entities. These enterprises, too, have become highly specialized operations. Complementing this development has been the creation of still another array of specialized off-the-farm functions—the manufacture of farm supplies, including implements, tractors, trucks, tractor fuels, fertilizers, feed supplements and mixed feeds, insecticides, and weed controllers, plus a host of other items. Today, the combined off-farm functions are considerably larger in magnitude than is the total operation of all our farms.

Despite the changes that have taken place, the functions of manufacturing farm supplies and storing, processing, and merchandising farm commodities still are closely related to agricultural production. Our farms could not operate for one week if they were cut off from these services. And, by the same token, the business firms which serve agriculture rely on farmers for their markets and for some of their supplies. There is a two-way interdependence with businessmen and farmers in the dual roles of suppliers and purchasers. Yet, in general, we tend to think of agriculture and business as separate entities. So true is this that our language contains no word to describe their interrelated functions. Presently, the only means of expressing such an idea is to write a paragraph or a page explaining it. Our vocabulary has not kept pace with progress.

To enable us to think more precisely in this field, the authors suggest a new word to describe the interrelated functions of agriculture and business—the term *agribusiness*.[1] By definition, agribusiness means the sum total of all operations involved in the manufacture and distribution of farm supplies; production operations on the farm; and the storage, processing, and distribution of farm commodities and items made from them. Thus, agribusiness essentially encompasses today the functions which the term agriculture denoted 150 years ago.[2]

[1] A word first used publicly by John H. Davis in a paper presented at the Boston Conference on Distribution, October 1955 (Boston, Retail Trade Board, copyright 1955).

[2] Since agribusiness is a new term, many of the component parts of this economic entity had to be defined for their particular use in this study. Therefore, the reader will find in Appendix I definitions of the basic terminology used.

Purpose of the Study

The purpose of this study is to present a concept of agribusiness and, with this as a frame of reference, to survey domestic food and fiber operations as a means of (1) contributing to a better understanding of existing relationships, particularly as between on-farm and off-farm functions; and (2) indicating an approach for improving policies relating to food and fiber. *The general hypothesis underlying the study is that so-called farm problems presently confronting the food and fiber segment of the economy are agribusiness rather than agricultural in nature and scope and, therefore, that such problems should be approached with agribusiness rather than agricultural perspective.* Also implicit is the supposition that the drift toward governmental farm support programs has been brought about, or at least greatly accentuated, by past failures to deal with so-called farm problems as agribusiness problems. From this supposition it would follow that the soundest approach to a healthy food and fiber economy with greater self-reliance on private effort lies in resolving so-called farm problems on an agribusiness basis through the cooperation of both farm and business leaders.

Scope and Methodology

This publication is directed primarily to an audience of professional and technical people interested in the problems of the food and fiber economy. This includes such groups as researchers, educators, farm and business leaders, and government policy makers. The authors recognize that a more general interest exists in the problems discussed in this study and anticipate that other publications of a less technical nature will be forthcoming.

This study is divided into four chapters. The first three describe the evolution of agribusiness, its dimensions and boundaries, and the economic activity within and outside of these boundaries, and the last considers the future of agribusiness.

The methodology employed is one of utilizing various types of data to measure the scope and amount of agribusiness interrelationships and intrarelationships. Rather than relying only on static equilibrium analysis, the authors have analyzed actual purchasing, production, and distribution patterns and income, sales, and capital positions to indicate dimensions and magnitudes of agribusiness and its component parts. The Leontief input-output method, which measures the flow of goods and services among and within industries, has been used to assist the reader in tracing such movements. The authors feel that in a changing, dynamic economy, agribusiness can best be explained by analyzing the actual flow of goods and services through economic entities.

The sources of material are the published and unpublished findings of Professors Leontief and Gilboy's Harvard Economic Research Project, the Department of Agriculture, the Department of Commerce, the Bureau of Labor Statistics, and various other agricultural and business publications.

Agribusiness and Economic Progress

In this study the authors attempt to solve no farm problem or problems. Instead, they undertake to portray agriculture in an agribusiness setting, against the background of the national economy. It is our belief that students and policy makers must examine agriculture and so-called farm problems with such perspective if they are to find sound and adequate solutions—solutions which will put to use wisely our capacity to produce food and fiber. Essentially the agribusiness approach, which is discussed in Chapter 4, is a method of examining old problems—the so-called farm problems—in a new and more comprehensive setting. In a modest way its contribution toward the solution of agricultural and agribusiness problems might be likened to that which the concept of the "Flow of National Income" has been to the solution of national problems during the past quarter century. Of itself it is not a program or even an answer to a specific problem. Rather, it offers a new frame of reference for old problems which can lead toward new and sounder policies and programs. The authors trust that this study may serve as a stimulus for more definitive studies within the agribusiness framework here set forth—studies which will contribute toward the solution of agricultural and agribusiness problems.

CHAPTER 1

The Genesis and Evolution of Agribusiness

Agricultural Self-Sufficiency

In the year 1800 it was appropriate to think of agriculture as more or less a self-contained industry, in that a farmer could operate successfully cut off from other industries. The typical farm family produced its own food, fuel, shelter, draft animals, feed, tools, and implements and even most of its clothing. Only a relatively few necessities had to be bartered for or purchased from off the farm. For the economy to operate on this basis required that 80% of our total labor force be engaged in farming; otherwise there would have been a deficiency of food and fiber to sustain the population.

The significant fact, in relation to this study, is that this type of agriculture involved the performance by the farm family of virtually all operations pertaining to the production, processing, storage, and distribution of farm commodities.

Technological Revolution on the Farm

Near the very end of the eighteenth century, the vanguard of a new technological era appeared on the agricultural horizon in the form of new mechanical devices designed to perform old tasks in less time and with less labor. Particularly prophetic of the spirit of this new era were such inventions as the cotton gin and the cast iron plow —both making their appearance in the 1790's.

During the first three decades of the nineteenth century, numerous inventions appeared, few of which were significant except as stepping stones to the future. In general, these devices proved cumbersome, fragile, or defective, with the result that few of them were put to general use. Then about 1830 agricultural mechanization gained new momentum—a characteristic it has retained ever since. With the settling of the virgin lands of the frontier and improved transportation, first by water and then by rail, new incentives appeared to speed the adoption of labor-saving devices. In the decade of the 1830's John Lane, John Deere, and William Parlin pioneered in the development of a steel mouldboard plow which was to revolutionize the process of preparing soil for planting. Simultaneously, McCormick, Manning, and Hussey concentrated on devices for cutting grass and grain by means of a reciprocating knife passing through bladed guards. This led the way to a rapid procession of improvements in the form of mowers and reapers.

Each important new development called forth others by creating bottlenecks in allied operations. Thus, the plow necessitated improved methods of harrowing, seeding, and cultivating. The reaper advanced the bottleneck of harvesting to the process of winnowing, thus creating the need for a mechanical thresher. But with each new need, inventors came forward to meet the challenge. By 1860 a host of new machines had appeared in the form of plows, harrows, planters, discs, stalk cutters, bailing presses, feed grinders, "straddle row" cultivators, mowers, reapers, and threshers.

Farmers, having tasted of the advantages and benefits of such new devices, became increasingly ready to try the next one. The task of improving old types of equipment and inventing new types has continued, gaining additional momentum and acceleration with each passing decade. The adaptation of the steam engine, the internal combustion engine, and the electric motor as sources of power on the farm has ushered in successive new eras of progress in terms of mechanized farming. Of all such developments, doubtless the internal combustion engine has proved most revolutionary in terms of farming methods. For not only has it made available a more powerful and convenient source of draft power; it has also made possible an array of companion implements, such as field cutters, loaders, ditchers, and harvesters, which would not have been feasible in a horsedrawn era.

Simultaneous to the mechanization of agriculture has been the application of research to other phases of farming, including plant and animal breeding, plant and animal nutrition, soil and water management, disease and insect control, animal housing, and commodity storage. Here, too, the rate of discovery has mounted decade by decade. Particularly notable in terms of advancing the scientific side of farming were the Morrill Act of 1862 creating land grant college systems; the establishment of the United States Department of Agriculture the same year; the Hatch Act of 1887 giving impetus to state experiment stations; the Smith-Lever Act of 1914 authorizing the Extension Service; and the Smith-Hughes Legislation of 1917 promoting vocational agriculture as a subject in secondary schools.

Technological Revolution off the Farm

Accompanying the technological revolution on the farm has been a companion revolution, related to food and fiber, off the farm. One important phase of this revolution has been the development of commercial facilities for handling, storing, processing, and distributing farm commodities and products made therefrom.

The textile industry led the way in this field, progressing for several decades even well ahead of farm mechanization. As early as 1800 the basic processes essential to the mechanical production of textiles were established. With this development, weaving in the home began to decline and then gradually disappeared. In general, off-farm processing of food developed much later than the industrial production of textiles—being closely related to the release of workers from agriculture and the migration of population to industrial centers.[1] This concentration of population in industrial centers created a corresponding need for moving food from the farm to urban areas. The magnitude of this feeding problem increased constantly with each decade. In 1957, agriculture employed about 12% of the working force in contrast to 72% in 1820 and 59% in 1860.

[1] The principal exception was the milling of grain, which was well advanced in colonial days. Early milling was done generally with stone burrs, driven by water power. Following the Civil War flour milling was revolutionized by the introduction of the moller mills method. This, together with the rapid expansion of wheat production in the West, led to larger milling units, concentrated at strategic locations.

Paralleling this change has been the development of a commercial food processing and distributing industry which not only has fed urban people as well as their farmer ancestors fared, but actually has improved the national diet in terms of quality, variety, and nutrition. In fact, today, the farm population itself depends in large measure upon the food industry for its daily needs. In other words, the modern farmer finds it more satisfactory to have industry process much of his own food than to process it on the farm.

To meet the food requirements of a constantly increasing urban population, a technological revolution has taken place with respect to methods of handling, preserving, and distributing food. Here, too, the tempo of progress has accelerated with passing years. Not only have older techniques such as drying, salting, canning, and preserving been improved, but new processes of dehydrating, concentrating, quick-freezing, pre-cooking, and handling fresh products have been developed. Related to these changes have been improvements in transport, refrigeration, grading, packaging, sanitation, etc. Simultaneously with these improvements has come the evolution of the food chain and the supermarkets as techniques for mass merchandizing. Also, the growth of public eating establishments to meet the needs of urban dwellers and to serve an increasingly mobile population has been significant.

A second phase of the off-the-farm technological revolution has evolved from the increasing tendency of farmers to utilize production supplies originating from off the farm. This tendency got an early impetus from agricultural mechanization. Farmers themselves had neither the means nor the skills to manufacture the more complex machines; hence they purchased them from industrial firms. Thus there developed alongside of agriculture a farm machinery industry which grew as mechanization advanced and as the country grew.

Supplementing this growth related businesses developed, such as binder twine factories, belt factories, implement stores, and repair shops. With the invention of the farm tractor, the petroleum industry expanded to meet the farmer's needs. The change-over from steel to rubber tires on tractors called on the rubber manufacturer for a new product. The electrifying of farm homes created need for a host of industrial items, such as motors, milking machines, milk coolers, pumps, waterers, heaters, welders, and power tools, all of which are a part of the modern farm.

As scientific knowledge as to the type and quality of soils improved, the farmer demanded more and better fertilizers to replenish the nutrients taken from his soil. Industry responded to this need with the rapid development of a commercial fertilizer industry. Also accompanying the improvement in farming methods came the incentive for better seeds and better livestock. This in turn led to specialized seed production and artificial insemination. To keep pace with improved livestock, farmers felt the need for better feeding practices, and industry responded by producing feed supplements, minerals, and mixed feeds. This tendency of farmers to shift to production supplies originating off the farm has grown with increasing tempo until today they are purchasing roughly half of all the supplies they use.

Agribusiness—A Product of the Dispersion of Functions

The significant point of the preceding discussion is this—during the past 150 years the food and fiber segment of our economy has evolved from a status of self-sufficiency to one of intricate interdependency with great segments of our industrial economy. Succinctly stated, it has evolved from an *agricultural* to an *agribusiness* status. In some *cases,* whole new industries have come into being to supply the needs of modern agriculture. Illustrative of these are, for instance, the farm implement companies, the meat packers, the food canners, and the food freezers. In other cases, established industries, such as rubber, chemicals, and pharmaceuticals, have expanded to create new products to meet the farmer's needs.

The present agribusiness economy has come about by the gradual dispersion of functions from agriculture to business, particularly those relating to the manufacture of production supplies and the processing and distribution of food and fiber products. This has continued to the point where today agriculture retains primarily the function of producing crops and livestock.

It is important to keep in mind that modern agribusiness is the result of a combination of forces actively at work for a century and a half and with roots running back even deeper into history. In no sense is it the result of a preconceived plan or design being carried to completion. Rather, it is the product of a complex of evolutionary forces more or less spontaneously at work without central guidance or direction. In fact, so gradual has been the development of agribusiness that students of agriculture and business largely have failed to recognize its significance.

Agribusiness has no center of control or direction. It has no president, no board of directors, and no central office. Instead, it consists of several million farm units and several thousand business units—each an independent entity, free to make its own decisions. In addition, there are hundreds of trade associations, commodity organizations, farm organizations, quasi-research bodies, conference bodies, and committees, each largely concentrating on its own interests. In general, these groups function in the area of education, promotion, coordination, and lobbying. They possess little or no direct power of determination over the business decisions of their members. The United States Government, also, is a part of agribusiness, to the degree that it engages in research, regulation of food and fiber operations, or the ownership and trading of farm commodities. The land grant colleges with their teaching, experiment stations, and extension functions also are an integral part of agribusiness. In brief, today agribusiness exists in a vast composite of decentralized entities, functions, and operations relating to food and fiber.

The evolution of agribusiness has brought with it innumerable benefits in the form of reduced drudgery, the release of workers for nonagricultural endeavors, better quality of foods and fibers, greater variety of products, improved nutrition, increased mobility of people, and so on. At the same time it has brought with it numerous problems of imbalance and maladjustments—problems which to a large extent reflect the uneveness of the evolutionary progress which has taken place. The complex problems, mentioned earlier, relating to commercial farming and low-income families are, to a large degree, of such origin.

Having examined the genesis and evolution of agribusiness, let us turn to a more detailed examination of its present general nature.

CHAPTER 2

The Nature of Agribusiness

As has been indicated, agribusiness comprises an important part of our economy—important both in size and in the fact that it supplies us with the very essential items of food and clothing. At the same time it is a complex, decentralized, segmented, and many-sided entity which is hard to visualize, describe, or analyze in concrete terms. Statistics describing it never have been pulled together in one place. Economists and analysts as well as farmers and businessmen tend to look upon farming and business as basically distinct and separate operations. This thinking is reflected in the make-up of national trade organizations, both general and commodity types; in the administrative arm of government with its separate Departments of Agriculture and Commerce; in Congress with its separate legislative committees; and in our educational system with its schools of business and agriculture largely isolated from one another.

In this chapter we shall attempt to examine the current general nature of agribusiness to determine its over-all magnitude, dimensions, and composition. This we shall do, first, by appraising it in terms of physical resources, working force, and the general flow of goods and services—both for agribusiness as a whole and in terms of its major components—and, secondly, by considering agribusiness in qualitative terms. When viewed in this manner, one will see agribusiness as a major component of our economy—one that comprises between 35% and 50% of the national total, depending on the type of yardstick one employs. Today, the larger portion of this food and fiber economy functions off the farm in terms of operations relating to the manufacture and handling of farm supplies and the processing, storage, and distribution of farm products. When surveyed as a whole, agribusiness will appear as an entity consisting of numerous functions which are more or less loosely tied together. Also apparent will be the susceptibility of the farming phase of agribusiness to a cost-price squeeze which has given rise to the creation of price support programs.

While the business firms and farm units comprising agribusiness could be grouped in various ways, the authors have combined them into three aggregates, Farm Supplies, Farming, and Processing-Distribution, or a composite called the Primary Agribusiness Triaggregate.[1] As the titles indicate, the first includes all operations pertaining to the off-farm manufacture, merchandising, and servicing of agricultural production supplies; the second includes operations on the farm; and the third includes off-farm activities relating to the conversion and merchandising of consumer items made from farm commodities.

Because available statistics have not been as complete or as uniform as one would desire, the authors have found it necessary to extrapolate, estimate, and otherwise adjust certain data in order to piece together a reasonably comprehensive picture of agribusiness. Since the purpose of this chapter is one of presenting only the general features and profile, however, these limitations do not present the handicap they would in a more detailed type of study. Both the sources of data and adjustments thereto will be indicated in the text or in footnotes as new material is introduced.

The Magnitude of Agribusiness

The General Nature of Agribusiness

Basic Dimensions. In order to get a comprehensive and composite picture of agribusiness and its component parts, we first shall examine the

[1] The Primary Agribusiness Triaggregate is used here to describe the three major parts of agribusiness. In a later section (see footnote 24, page 31), a Secondary Agribusiness Triaggregate is discussed and is composed of the Farming Aggregate and a breakdown of the Processing-Distribution Aggregate into Food Processing and Fiber Processing. The Farm Supplies Aggregate is excluded in the Secondary Agribusiness Triaggregate.

Agribusiness Flow Chart for 1954 (Exhibit 1).[2] In doing this we shall, in effect, be skipping somewhat ahead in the "story" and then executing a "flashback" in this and the following chapter for the purpose of explaining the method of analysis and the significance of various interrelationships.

From this flow chart one can note that in 1954 consumers spent about $93 billion[3] for the end products and services relating to food and fiber. Essentially this represents the total bill paid by consumers for all production, processing, fabricating, and distribution of such end products. Although the aggregate consumer expenditure is in terms of purchasers' values, all the remaining items are in producers' values.

Of this $93 billion consumer item, about $75 billion related to commodities grown on American farms, $15 billion to wholesale and retail price margins, and $3 billion to all other items, including fish, synthetic fibers, and imported foods and fibers.[4] In the production, processing, and distribution of the $93 billion of end products from food and fibers, the following transactions took place at the functional levels indicated:[5]

	In billions
Purchases of manufactured production supplies by farmers	$16.4[6]
Sales of commodities by farmers	29.6
Sales of food and fiber end products by processors	69.9
Purchases by consumers	93.0

These transactions will be analyzed in more detail in Chapter 3. The important fact to note here is that this $93 billion spent by consumers for food and fiber items constitutes roughly 40% of the $236.5 billion total consumer expenditures for the year 1954.[7] These figures of $93 billion and 40% serve as yardsticks for measuring the general size and massiveness of agribusiness against the background of the national economy. Even though today food and fiber items occupy only about half as important a position in our economy as they did 150 years ago, they still are a significant part of our economy when both on-farm and off-farm operations are included.

As indicated in the tabulation that follows, the dollar volume of food and fiber end products purchased by consumers increased about $21 billion from 1947 to 1954. Even so, such purchases constituted a relatively smaller proportion of total consumer purchases in 1954 than in 1947.

Consumer Purchases of Agribusiness Items in Billions of Dollars and in % of Total U.S. Expenditures for Personal Consumption[8]

	1947		*1954*	
Food	$38.7	23.4%	$55.4	23.4%
Meals and Beverages	$11.9	7.2%	$14.4	6.1%
Subtotals	$50.6	30.6%	$69.8	29.5%
Tobacco Products	$ 3.9	2.3%	$ 5.3	2.2%
Shoes and Footwear	$ 3.0	1.8%	$ 3.5	1.5%
Clothing and Accessories	$15.6	9.5%	$16.0	6.8%
Subtotals	$22.5	13.6%	$24.8	10.5%
Totals	$73.1	44.2%	$94.6[9]	40.0%

[2] The figures shown in the 1954 Agribusiness Flow Chart were derived by the authors with the cooperation of the United States Department of Agriculture by extrapolation from the 1947 Agribusiness Flow Chart (Exhibit 3, Chapter 3), using more current data. The 1947 Agribusiness Flow Chart, in turn, was developed from estimates of the U.S. Department of Agriculture and from the 1947 input-output study of direct purchases to and sales from specific industry sectors. The input-output study is based on 1947 census data and was originally put together by the Bureau of Labor Statistics under the direction of Harvard economist, Professor Wassily Leontief. A more detailed explanation of the sources of the data and the method of selecting the industries included in both the 1947 and 1954 Agribusiness Flow Charts will be found in footnote 21 on page 29 in Chapter 3 of this study.

[3] Here and in the text that follows the authors have resorted to the use of three estimates of total consumer expenditures, ranging from $93 billion to $95 billion; one each from the U.S. Department of Agriculture, the U.S. Department of Commerce, and *The Economic Almanac*. This policy has been pursued because no one source provided sufficient detail for the analysis desired and because of the impracticability of reconciling the figures and their supporting data. The figure shown here was obtained from unpublished data of the U.S. Department of Agriculture (see footnote 2, above). The estimates of retail and wholesale price margins and food and fiber imports were also obtained from the U.S. Department of Agriculture (hereafter referred to as U.S.D.A.).

[4] See footnote 2, above.

[5] Of course, the dollar value of transactions would vary with the detail of the breakdown of functions.

[6] Includes some of the wholesale and retail margins paid by farmers for their purchased inputs.

[7] See *Survey of Current Business*, February 1956, p. 9.

[8] The Conference Board, *The Economic Almanac* (New York, Thomas Y. Crowell Company, 1956).

[9] The $94.6 billion is used here only to demonstrate the breakdown of food and fiber expendiures even though it cannot be reconciled with our earlier $93.0 billion figure (see footnote 3, above).

EXHIBIT 1. AGRIBUSINESS FLOW CHART: 1954

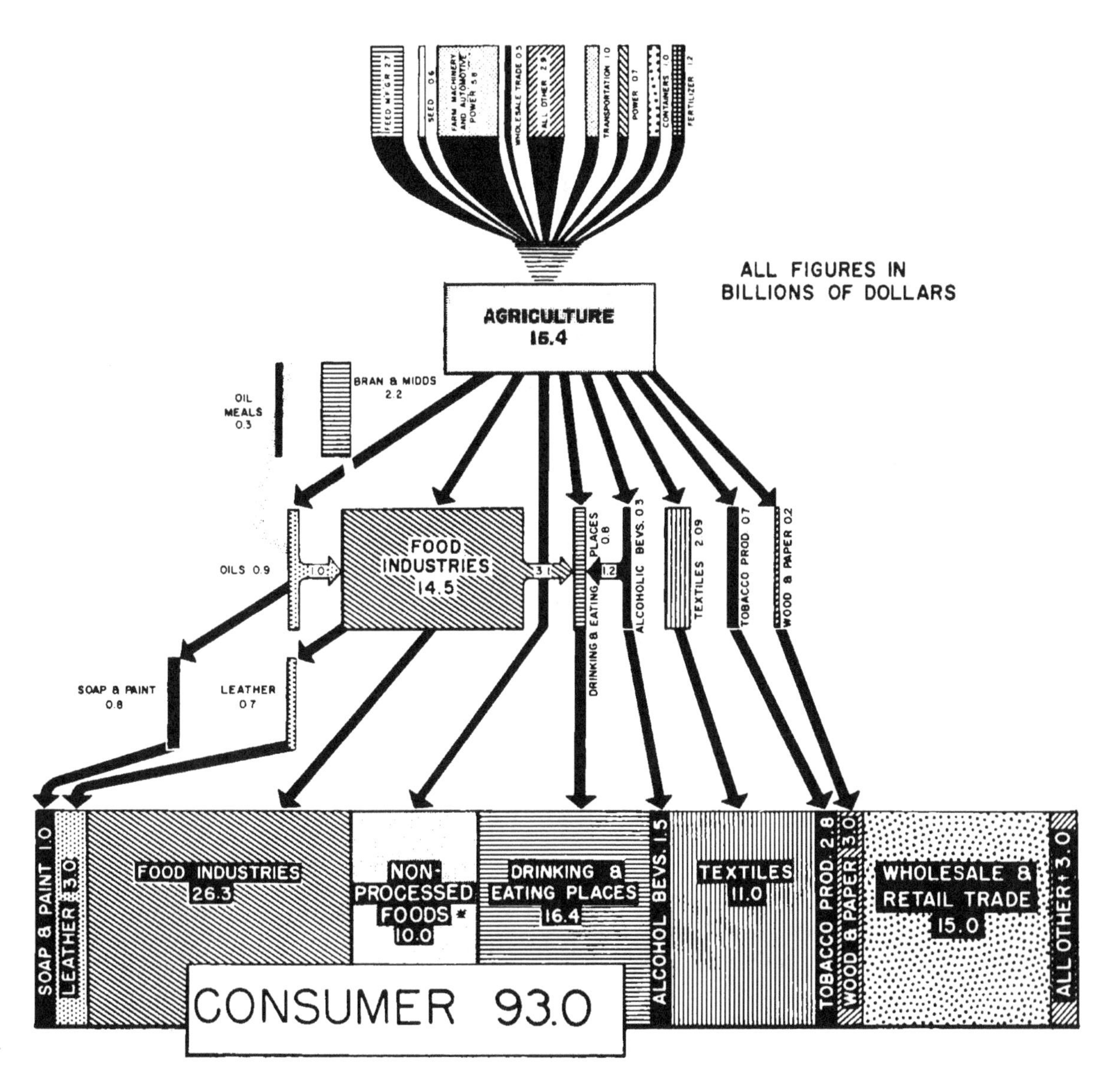

* Includes food products purchased by consumers in relatively unchanged form (like eggs or fresh vegetables) and also those consumed on the farm.

† Includes sea foods, synthetic fibers, and imports.

SOURCE: See footnote 2, page 8.

TABLE 1. SELECTED BALANCE SHEET DATA RELATING TO AGRIBUSINESS: DECEMBER 31, 1947, 1954
(Billions of dollars)

Assets	*1947*	*1954*	*Liabilities*	*1947*	*1954*
Inventories	$22.2	$28.1	Current Liabilities	$14.3	$23.9
Other Current Assets	32.8	39.2	Noncurrent Liabilities	8.5	14.5
Total Current Assets	55.0	67.3	Total Liabilities	22.8	38.4
Fixed Assets	112.5	149.6	Net Worth	147.0	181.8
Other Noncurrent Assets	2.3	3.3			
Total Noncurrent Assets	114.8	152.9			
Total Assets	$169.8	$220.2	Total Liabilities and Net Worth	$169.8	$220.2

NOTE: The balance sheet data shown here were derived from the U.S.D.A. Balance Sheet of Agriculture and the reports of the Federal Trade Commission and Securities Exchange Commission on United States Manufacturing, Retail and Wholesale Corporations.

In the case of business firms included in agribusiness, these estimates tend to understate the true picture because they include only incorporated firms. (The resulting underestimate of the assets of the nonfarm phases of agribusiness may amount to as much as 10%.) In contrast, the assets of the Farming Aggregate tend to be on a relatively higher base since U.S.D.A. data reflect a reasonable market value of real property and because estimates include the value of farm dwellings. The authors have classified farm lands and buildings as physical plant and equipment, and crops held by farmers as inventory even when pledged to the Commodity Credit Corporation as collateral. Also, livestock have been classified as between "plant" and "inventory" on the basis of age, in line with U.S.D.A. practices.

A still more difficult problem arises with respect to those firms engaged only partially in agribusiness—particularly with respect to the manufacturers of farm supplies. This the authors have attempted to resolve by prorating the balance sheet items of the respective industries on the basis of the relative volume of end products used on farms. The factors used were developed in part from the relationships set forth in the Harvard Interindustry Study for 1947 and in part from unpublished data prepared by the Food and Materials Requirements Division, Commodity Stabilization Service, U.S.D.A., for use in determining raw material needs of agriculture in the event of national emergency or war. The factors used in prorating industrial assets to agribusiness are as follows:

Manufacturing Industries		*Wholesale Industries*		*Retail Industries*	
Chemical	10%	Motor vehicles and automobiles	15%	Building material and farm equipment	70%
Petroleum	20	Chemicals	10	Auto dealers and gas service stations	10
Rubber	20	Machinery equipment and supplies	8		
Iron and steel	5				
Machinery	5				
Motor vehicles and parts	15				

Still another problem grew out of the fact that adequate data for wholesaling and retailing operations relating to food and fiber were not available for the years 1947 and 1954. Therefore, the authors have extrapolated, applying the 1947 and 1954 wholesale indices to the 1950 and 1951 balance sheets of wholesale and retail corporations, respectively.

Capital Structure. Another facet of agribusiness is revealed by an examination of the assets it employs. In any attempt to construct a balance sheet for agribusiness, a number of problems are encountered. Complete data are not available for all components; land and buildings in the case of agriculture are carried at market value rather than at the lower of market price or cost; farm assets include the value of buildings used as dwellings, and certain facilities such as tire plants and refineries are utilized only partially in serving agribusiness, thus making necessary a proration of assets. The methods used in dealing with these problems are explained in the footnote accompanying Table 1. Even recognizing the shortcomings of the figures in Table 1, however, these data indicate the magnitude, dimensions, and composition of one aspect of agribusiness.

If one applies the conventional methods of analysis to this improvised balance sheet of agribusiness, the result is quite a strong showing, as indicated in the accompanying tabulation. Even in 1954 the current ratio was almost 3 to 1 and that of equity to total liabilities was almost 5 to 1. The slight decrease in equity as a percentage of total assets from 86.6% to 82.6%, the decline in the percentage of current assets from 32.4% to 30.6%, and the slight rise in noncurrent liabilities from 37.3% to 37.8% all reflect a less liquid financial structure in agribusiness in 1954 than in 1947.

	1947	*1954*
Current ratio (current assets to current liabilities)	3.85	2.82
Equity to total liabilities (ratio)	6.45	4.73
Equity as % of total assets	86.6%	82.6%
Current assets as % of total assets	32.4%	30.6%
Noncurrent liabilities as % of total liabilities	37.3%	37.8%

Further insight into the capital structure of agribusiness is provided in the tabulation below which relates its balance sheet totals to those of All U.S. Manufacturing, Wholesale and Retail Corporations (including industries making farm supplies and processing and distributing farm products), and the Agricultural Balance Sheet. Even allowing for a considerable margin of error in the estimates of agribusiness balance sheet items, it is apparent that the capital requirements of agribusiness are high relative to those of American manufacturing industries generally. Agribusiness, possessing in excess of 50% of the combined assets of manufacturing industries and agriculture, accounts for only 40% of total U.S. consumer expenditures.

Manufacturing, Retail and Wholesale Corporations, and Agriculture, December 31, 1954
(Billions of dollars)

	Total	*Agribusiness*	*All Other*
Total assets	$386.4	$220.2	$166.2
Liabilities	96.7	38.4	58.3
Net worth	289.7	181.8	107.9

Labor Force. The nature of agribusiness can be seen still more comprehensively if one examines the working force. The figures that follow depict the agribusiness labor force in 1947 and 1954—relating it to that of the national economy.[10] In 1954 about 24 million persons out of a total working force of 64.5 million were engaged in agribusiness activities. While the number of workers in agribusiness remained almost unchanged from 1947 to 1954, the relationship of this number to the national working force decreased from 41% to 37%. When one relates this trend to the increase in consumer purchases of agribusiness products

	1947	*1954*
Agribusiness labor force (in millions)	24.5	24
National labor force (in millions)	60.2	64.5
Agribusiness labor force as % of civilian labor force	41%	37%
Average investment per worker in agribusiness[11]	$6,890	$9,175
Average investment per worker in "All Manufacturing Industries"	$9,320	$9,525

[10] Derived from *The Economic Almanac, 1956,* p. 353, and from estimates of the U.S.D.A.

[11] Most of the increase in the average investment (assets) per worker in agribusiness was due to the increase in assets per worker in agriculture from 1947 to 1954. (See page 14.)

between 1947 and 1954, it is apparent that a marked increase took place in terms of output per worker. Doubtless the 33% increase in average investment per worker in agribusiness from 1947 to 1954, indicated in the same tabulation, was a major contributing factor in bringing about this rise in output per worker. Also accompanying this development was a 24% increase in the average hourly earnings of workers in agribusiness, as may be seen from the following itemization. This 24% increase, however, compares with a 46% average wage increase for All Manufacturing Industries during the same period.

	Average Hourly Earnings		*% Change*
	1947	*1954*	
Agribusiness[12]	$1.17	$1.45	23.9%
All Manufacturing Industries[13]	1.24	1.81	46.0%

Income. Still another important aspect of agribusiness which further serves to portray its nature is its income position. It is difficult to get a true picture of the income status of agribusiness, however, because available data for agriculture include in net income not only entrepreneurial remuneration but also all compensation to the farm family for its own labor and investment. Further complicating the question of income to the agricultural phase of agribusiness are such issues as evaluating perquisites, the proper treatment of off-farm income, differentiating between commercial farmers and rural residents employed off-farm, and farmer compensation derived from government programs.

While statisticians have attempted to devise means of coping with some of these problems, it is still difficult to adjust farm income statistics for 1947 and 1954 so as to make them additive with those of other phases of agribusiness. For this reason no attempt is made here to present statistically an over-all operating income or net income picture for agribusiness as a whole. The analysis of the income status of its component aggregates in the following section of this chapter, however, will shed light on the income status of total agribusiness.

[12] Weighted average estimate includes farm income (minus 4¾% for capital invested as farm wages. See Table 5).

[13] Includes Agribusiness Manufacturing Industries.

TABLE 2. SUMMARY BALANCE SHEET OF AGRIBUSINESS AND ITS COMPONENT AGGREGATES: DECEMBER 31, 1947, 1954
(Billions of dollars)

	1947				1954			
Assets	*Suppliers of Agricultural Products*	*Farming Aggregate*	*Processing and Distributing Industries*	*Total*	*Suppliers of Agricultural Products*	*Farming Aggregate*	*Processing and Distributing Industries*	*Total*
Inventories	$2.1	$9.0	$11.1	$22.2	$3.2	$9.6	$15.3	$28.1
Other Current Assets*	2.6	20.5	9.7	32.8	4.3	22.0	12.9	39.2
Total Current Assets	(4.7)	(29.5)	(20.8)	(55.0)	(7.5)	(31.6)	(28.2)	(67.3)
Fixed Assets	2.7	102.5	7.3	112.5	5.6	131.5	12.5	149.6
Other Noncurrent Assets	0.6	—	1.7	2.3	1.0	—	2.3	3.3
Total Assets	$8.0	$132.0	$29.8	$169.8	$14.1	$163.1	$43.0	$220.2
Liabilities								
Total Current Liabilities	$1.8	$4.2	$8.3	$14.3	$2.9	$9.8	$11.2	$23.9
Long-term Debt and Other Noncurrent Liabilities	0.7	5.1	2.7	8.5	1.7	8.2	4.6	14.5
Total Liabilities	(2.5)	(9.3)	(11.0)	(22.8)	(4.6)	(18.0)	(15.8)	(38.4)
Proprietors' or Stockholders' Equity	5.5	122.7	18.8	147.0	9.5	145.1	27.2	181.8
Total Liabilities and Equity	$8.0	$132.0	$29.8	$169.8	$14.1	$163.1	$43.0	$220.2

* Other current assets for the suppliers of agricultural products and for processing and distributing industries include the following items: cash, U.S. Government securities (including Treasury savings notes and receivables from the U.S. Government excluding tax credits), other notes and accounts receivable, and miscellaneous current assets. Other current assets for the Farming Aggregate include the following items: deposits in currency, U.S. Savings Bonds, and investments in cooperatives.

Nature of Agribusiness Aggregates

The general characteristics of agribusiness, noted in the preceding section, in reality were the sum of the characteristics of the component aggregates of our food and fiber economy. Hence an examination of these components will facilitate a clearer understanding of the nature of agribusiness, in terms of both general properties and interrelationships.

Capital Structure. Table 2 presents a summary balance sheet with a breakdown for the Primary Agribusiness Triaggregate—Farm Supplies, Farming, and Processing-Distribution. These data, of course, are subject to the limitations described in connection with Table 1. Even so, they provide a general picture of the capital structure of agribusiness.

A glance at the following ratios derived from Table 2 will show the sound position of all three aggregates when measured by conventional balance sheet standards. Particularly noteworthy is the high ratio of equity to total liability and the high relationship of equity to total assets in the Farming Aggregate, reflecting in large measure

	Farm Supplies	*Farming*	*Processing-Distribution*
	1954		
Current ratio	2.6	3.2	2.5
Equity to total liabilities (ratio)	2.1	8.1	1.7
Equity as % of total assets	67.4%	89.0%	63.3%
Noncurrent liabilities as % of total liabilities	37.0%	45.6%	29.1%
	1947		
Current ratio	2.6	7.0	2.5
Equity to total liabilities (ratio)	2.2	13.2	1.7
Equity as % of total assets	68.8%	93.0%	63.1%
Noncurrent liabilities as % of total liabilities	28.0%	54.8%	24.5%

the large debt retirement by farmers during the war and immediate postwar years. Also significant are the downward trends in current ratio and equity to total liability ratio of the Farming Aggregate between the dates under consideration. These, doubtless, result from lower farm prices on the one hand and increased purchase of equipment and other facilities on the other.[14] The principal change affecting the Farm Supplies and Processing-Distribution Aggregates was the increase in

[14] Certain types of farm equipment still remained in scarce supply in 1947.

TABLE 3. ESTIMATED VALUE OF PRINCIPAL INPUTS PURCHASED BY FARMERS: 1947–1954
(Millions of dollars)

	Current Farm Operating Expenses, excluding Labor						*Farm Expenditures for Machinery and Depreciation on Machinery*	
Year	*Feed*	*Livestock*	*Seed**	*Fertilizer and Lime*	*Repairs and Operation of Motor Vehicles and Other Farm Machinery†*	*Miscellaneous‡*	*Capital Farm Machinery Expenses§*	*Depreciation of Farm Machinery‖*
1947	$3,746	$1,379	$514	$755	$1,776	$1,251	$1,901	$933
1948	3,996	1,589	581	826	2,116	1,431	2,662	1,293
1949	3,024	1,528	543	895	2,266	1,569	3,046	1,683
1950	3,330	2,000	531	978	2,340	1,611	2,931	1,935
1951	4,168	2,443	561	1,085	2,592	1,943	3,192	2,257
1952	4,302	1,917	594	1,229	2,717	1,945	2,768	2,459
1953	3,755	1,320	560	1,246	2,754	1,937	2,861	2,534
1954	3,872	1,563	557	1,274	2,628	1,896	2,649	2,567
Index of Change 1954/1947	103.4%	113.3%	108.4%	168.7%	148.0%	151.6%	139.3%	275.1%

* Includes bulbs, plants, and trees.

† Includes all gasoline and other petroleum fuel and oil used in the farm business as well as 40% of total operating costs for automobiles, exclusive of gasoline and oil (50% used in 1942–1945 figures).

‡ Includes short-term interest, pesticides, ginning, electricity and telephones (business share), livestock marketing charges, containers, irrigation, grazing, binding materials, tolls for sirup, horses and mules, harness and saddlery, blacksmithing and hardware, veterinary services and medicines, net insurance premiums, and miscellaneous dairy, nursery, greenhouse, apiary, and other supplies.

§ Includes expenditures for tractors, trucks (and for farm business use 40% of total farm purchases of automobiles, 50% in 1942–1945), and all other machinery and equipment with the exception of harness and saddlery items, etc., included in the "miscellaneous" category above.

‖ Depreciation in terms of current replacement cost—40% of total depreciation of farm automobiles (50% in 1942–1945).

SOURCE: *The Farm Income Situation*, July 17, 1956, Agricultural Marketing Service, U.S.D.A.

noncurrent liabilities as a percentage of total liabilities, reflecting the expansion of facilities. Careful examination of Table 2 reveals a very large fixed asset position for the Farming Aggregate—this aggregate accounting for more than half of the total fixed assets of all agribusiness. This is still true even if one discounts generously for the higher evaluation methods used with respect to assessing farm assets. Also noteworthy is the fact that the total assets of the Farm Supplies Aggregate almost doubled between 1947 and 1954. This increased capacity of the farm supplies industry to produce farm machinery and other equipment in turn is reflected in a $12 billion increase in the value of on-farm machinery between 1947 and 1954.[15]

This growth of the Farm Supplies Aggregate may be illustrated by the growth of the dollar inputs from this aggregate purchased by farmers from 1947 through 1954 (Table 3). From this, it appears that the dollar purchases of the various manufactured farm supplies increased in varying amounts between 1947 and 1954, thus confirming the previously observed increasing dependency of farming upon purchased inputs. Supplementary data from the U.S. Department of Agriculture indicated that between these dates the number of tractors on farms rose from 2.2 million to 4.6 million and the number of grain combines and corn pickers tripled.[16] The same report predicted that by 1960 the mere replacement of farm equipment would require an annual output on the part of the equipment industry equal to 95% of its total output in 1952. The Federal Reserve Board reported that $17.7 billion of farm machines were in use on farms as of January 1, 1955.[17] This amount, according to *Fortune* magazine,[18] was $10 billion

[15] *Statistical Abstract 1955*, p. 647.

[16] *Study of Farm Machinery and Equipment Replacement Needs*, U.S.D.A., May 1955.

[17] *Federal Reserve Bulletin*, July 1955, p. 873.

[18] *Fortune*, May 1956, p. 124.

more than the total net assets of the American steel industry, seven times that of the nonferrous metals industry, and five times that of the automobile industry.

Labor Force. Another facet of agribusiness is revealed when one examines the labor force of the aggregates. As indicated by the accompanying tabulation,[19] the changes in the size of the labor force of the respective aggregates between 1947 and 1954 were such as largely to offset one another, with the result that the total working force of agribusiness remained almost unchanged—about 24 million persons. Within agribusiness, however, the change in working force was of considerable magnitude, the Farming Aggregate losing 2 million workers and the Farm Supplies and Processing-Distribution Aggregates gaining 1 million and ½ million, respectively. This means that in 1954 the total agribusiness working force was distributed among aggregates: 25% in Farm Supplies; 33.3% in Farming; and 41.7% in Processing-Distribution. Thus, in our food and fiber economy two persons were employed off-farm for each one on-farm. These facts also would appear to harmonize with the previously noted heavy capital expansion of the Farm Supplies and Farming Aggregates—the former making possible the increased output of farm supplies and the latter a reduction in the farm labor force without decreasing the volume of total farm output. Further explaining the reduction in the number of farm workers is the increase in assets per worker in agriculture from $13,200 in 1947 to $20,400 in 1954.[20]

Number of Workers (Millions)

	1947	*1954*	*% Change*
Farm Supplies	5.0	6.0	20.0%
Farming	10.0	8.0	—20.0
Proc.-Dist.	9.5	10.0	5.3
Total Agribusiness	24.5	24.0	— 2.0%

Aggregate as % of Total Agribusiness

	1947	*1954*
Farm Supplies	20.4%	25.0%
Farming	40.8	33.3
Proc.-Dist.	38.8	41.7
Total Agribusiness	100.0%	100.0%

[19] Estimated by U.S.D.A.

Output and Income. Still another facet of agribusiness may be seen from an examination of the tabulation that follows setting forth the change in output and increased value per aggregate of agribusiness for the years 1910, 1947, and 1954.[21] Particularly significant is the rapid expansion of both the Farm Supplies and Processing-Distribution Aggregates in comparison with the Farming Aggregate, as indicated in the columns headed "% of Total Increased Value."[22] Even though farm output has expanded at a rate sufficient to feed and clothe our growing population at an ever-improving standard of living, the relative position of the Farming Aggregate, in terms of increased value, has decreased from 54% in 1910 to 17% in 1954. During the same period the increased value of the Farm Supplies and Processing-Distribution Aggregates has almost doubled—increasing from 46% to 83%. Accompanying this change has been the revolution in farming

	Value of Output (Billions)	*Increased Value* (Billions)	*% of Total Increased Value*
	1910		
Farm Supplies	$1.0	$1.0	11%
Farming	5.8	4.8	54
Proc.-Dist.	8.9	3.1	35
Total		$8.9	100%
	1947		
Farm Supplies	$12.8	$12.8	20%
Farming	29.3	16.5	26
Proc.-Dist.	62.9	33.6	54
Total		$62.9	100%
	1954		
Farm Supplies	$16.4	$16.4	21%
Farming	29.5	13.1	17
Proc.-Dist.	75.0	45.5	62
Total		$75.0	100%

[20] Derived from balance sheet assets and total farm labor force including farm owner-operators.

[21] *Historical Statistics of the U.S. 1789–1945*, U.S.D.A., 1949, pp. 99, 100, 183, and Figures 1 and 3.

[22] All amounts are in producers' values with the exception of the Farm Supplies Aggregate which includes some retail and wholesale price margins.

efficiency and in the variety, quality, and merchandising service related to food and fiber products, discussed in Chapter 1.

Still other characteristics of agribusiness are brought into focus as one examines the net income picture of the component aggregates. The income picture of the Farm Supplies Aggregate is summarized in the following tabulation:[23]

	1947	*1954*	*% Change*
	(Billions)		
Net sales	$12.9	$20.1	55.8%
Cost of goods and expenses	11.8	18.5	56.8
Net profit from operations	1.1	1.6	45.5
Other income or deductions (net)	0.1	0.1	0.0
Net profit before federal income taxes	1.2	1.7	41.7
Provisions for federal income taxes	0.4	0.7	75.0
Net profit after taxes	0.8	1.0	25.0
Net profit after taxes as % of net sales	6.2%	5.0%	—19.4%

The net sales of this aggregate, in dollar terms, was 56% greater in 1954 than in 1947, further reflecting the marked increase in the use of purchased inputs by farmers in recent years. This provides an average annual gain in net sales of about 8%, which compares favorably with the average rate of growth of American industry as a whole during the same period. Simultaneously net profits from operations rose by 45%, net profits before taxes by 41.7%, and net profits after taxes by 25%. However, when net profits are related to net sales, the rate of earning was 19% lower in 1954 than in 1947.

For reasons explained earlier in this chapter, it is not possible to analyze net farm income in terms of the conventional methods commonly applied to business income. Nor is it feasible to adjust farm income data to the business basis by means of imputed cost factors or by extrapolation. A reasonably realistic picture of the income status of the Farming Aggregate, however, may be obtained from a study of Table 4. Even though these data are not in the form most desirable for our purpose, still a comparison of the two years does have validity, particularly since the method of analysis was the same for both periods. Especially striking is the fact that even though total gross farm income increased by 1.7% between 1947 and 1954, residual income to farm operator families decreased by 26.3%. (This residual income includes compensation for entrepreneurial function, family labor, and family investment in operations.) Primarily responsible for this decrease in residual income was the 41.4% rise in total farm production costs—the major contributory factor being increases of 111.8% in depreciation and maintenance, 70.5% in cost of vehicle operation, 59.3% in real and personal taxes, 57.5% in expenditures for fertilizer and lime, and 35.2% in outlay for seeds. Because cash receipts from marketings were almost identical for both years (0.2% lower in 1954), the change in the ratio of residual income to cash receipts from marketing coincides closely with the change in residual income—a drop of 26.3% and 26.5%, respectively.

Further insight into the income status of the Farming Aggregate can be gained from Table 5 which attempts to depict residual income in terms of returns per hour to farm labor, both family and hired.[24] This analysis is made on the basis of the estimated number of manhours required to produce a given year's output, assuming average adult male workers performed the task, rather than on the basis of the actual number of workers engaged in agriculture times the actual hours devoted to the task. Use of this method avoids the necessity of adjusting data for part-time workers and of evaluating the efficiency factor for each type of worker.

When thus calculated, the "Realized Return per Hour to All Farm Labor and Management" dropped from $.98 in 1947 to $.73 in 1954[25]—a decline of 25.5%. This is in contrast to an average hourly wage rate for all industrial workers of

[23] Derived from Table A-1, Appendix II.

[24] The left-hand column in Table 5, "Total Realized Return to All Farm Labor and Capital," can be reconciled with the item, "Residual Income to Farm Operators (Families)," in Table 4, by adjusting the latter to include wages to hired farm labor, farm mortgage interest, rent to nonfarm landlords, and short-term interest.

[25] "Realized Return per Hour to All Farm Labor and Management" declined to an estimated $.64 in 1955. Note that these returns per hour have been computed on the basis of the number of manhours required for an average adult male worker to perform the respective farm jobs. If actual labor numbers on the farm were used, the return per hour would be much lower because of the underemployed labor force in agriculture.

TABLE 4. OPERATING DATA FOR FARMING AGGREGATE: 1947, 1954
(Millions of dollars)

	1947	*1954*	*Percentage Change*
Total Gross Farm Income:			
Cash Receipts from Farm Marketings	$30,014	$29,954	— 0.2%
Government Payments to Farmers	314	257	—18.2
Home Consumption of Farm Products	3,095	1,895	—38.8
Rental Value of Farm Dwellings	1,220	1,741	42.7
Net Change in Inventory	—1,059	318	
Total	$33,584	$34,165	1.7%
Production Costs:			
Feed Bought	$3,746	$3,800	1.4%
Livestock Bought (except Horses and Mules)	1,416	1,483	4.7
Fertilizer and Lime Bought	746	1,175	57.5
Vehicle Operation	1,305	2,225	70.5
Depreciation and Maintenance	2,338	4,951	111.8
Taxes on Farm Real Estate and Personal Property	705	1,123	59.3
Seed Bought	514	565	9.9
Miscellaneous	1,546	2,090	35.2
Total	$12,316	$17,412	41.4%
Expense:			
Wages to Hired Labor (Cash and Perquisites) *	$2,837	$2,985	5.2%
Net Rent†	1,474	1,086	—26.3
Interest to Holders of Farm Mortgages	222	380	71.2
Total	$4,533	$4,451	—1.8%
Total Production Costs and Expense	$16,849	$21,863	29.8%
Residual Income to Farm Operators (Families)	$16,735	$12,302	—26.5%
Residual Income as a Percentage of Cash Receipts from Marketing	55.8%	41.1%	—26.3%

* Does not include proprietors' wages or farm wages of other members of the family.
† Net rent also contains that part of government payments (included as income above) that is paid to landlords not living on farms.
SOURCE: U.S.D.A., Agricultural Research Service, *The Balance Sheet of Agriculture* (Washington, January 1, 1948, and January 1, 1955).

$1.24 in 1947 and $1.81 in 1954—an increase of 46.0%. Farm and industrial wage rates are not completely comparable since the food produced on the farm is valued at farm prices and the housing at rural rental rates in the case of the return to farm labor. If one assumes that the rural value of these factors is half that of urban areas (which appears realistic) and therefore doubles the income items of "Home Consumption of Farm Products" and "Rental Value of Farm Dwellings" (see Table 4) and adjusts the figures in column 1, Table 5, for 1947 and 1954 accordingly, the resulting "Realized Return per Hour to All Farm Labor and Management" becomes $1.23 and $.97 for 1947 and 1954, respectively—a decline of 21% during the period. Even on this basis the resulting estimated return to farm labor for 1954 would be only about half the average wage paid industrial workers.

Also noteworthy in column 4, Table 5, is the 2.9 billion decline between 1947 and 1954 in estimated manhours required for farm production—a reduction of 16%. Actually the increase in worker productivity was somewhat greater than these figures would indicate because the total farm output was about 9% greater in 1954 than in 1947. This average increase in output per farm worker per hour per year of approximately 4% compares favorably with that of American industry for the same period.

On the whole the ability and skill required in farming would seem to match that required in modern industry. On the typical family farm the number of skills and the scope of knowledge required per worker are even greater than in industry. Should one include in the type of analysis contained in Table 5 an imputed hourly wage equal to the average hourly industrial wage for

TABLE 5. ESTIMATED RETURN PER HOUR TO ALL FARM LABOR: 1929–1954

Year	*Total Realized Return to All Farm Labor and Capital* (Millions of Dollars)*	*Allowance for Capital at 4¾%† (Millions of Dollars)*	*Net Return to Labor and Management (Millions of Dollars)*	*Total Manhours Required for Agricultural Production‡ (Millions of Hours)*	*Realized Return per Hour to All Farm Labor and Management (Dollars)*
1929	$8,982	$2,998	$5,984	23,158	$.258
1930	6,908	2,955	3,953	22,921	.172
1931	4,775	2,598	2,177	23,427	.093
1932	3,419	2,174	1,245	22,605	.055
1933	4,209	1,809	2,400	22,554	.106
1934	5,407	1,914	3,493	20,232	.173
1935	6,261	1,984	4,277	21,052	.203
1936	6,889	2,127	4,762	20,440	.233
1937	7,080	2,197	4,883	22,097	.221
1938	6,047	2,198	3,849	20,577	.187
1939	6,247	2,136	4,111	20,680	.199
1940	6,245	2,157	4,088	20,443	.200
1941	8,553	2,250	6,303	20,049	.314
1942	11,831	2,584	9,247	20,849	.444
1943	15,386	3,038	12,348	20,682	.597
1944	15,871	3,434	12,437	20,482	.607
1945	16,604	3,765	12,839	19,108	.672
1946	19,303	4,086	15,217	18,423	.826
1947	21,849	4,577	17,272	17,593	.982
1948	20,765	5,056	15,709	17,116	.918
1949	18,175	5,306	12,869	16,563	.777
1950	17,367	5,197	12,170	15,227	.799
1951	19,713	6,008	13,705	15,632	.877
1952	19,170	6,618	12,552	15,241	.824
1953	18,405	6,364	12,041	15,103	.797
1954	16,778	6,068	10,710	14,738	.727

* Includes realized net income of farm operators, wages to hired farm labor, farm mortgage interest, rent to nonfarm landlords, short-term interest, and foods produced and consumed on the farm.
† Allowance for capital is computed at 4¾% of current value of farm real estate, inventory value of crops and livestock, inventory value of motor vehicles and machinery (excluding 60% of the automobile), and an allowance for working capital. This rate approximates the interest rate on farm-mortgage debt in recent years.
‡ Labor requirements in terms of the number of manhours required for an average adult male worker to perform the various farm jobs.
SOURCE: U.S.D.A., Agricultural Marketing Service.

1947 and 1954, the Farming Aggregate would have shown a deficit of about $2 billion in 1947 and $11 billion in 1954.[26] While these data do not present such a clear picture of the income status of the Farming Aggregate as might be desired, they do indicate a marked decline in its relative net income position.

The over-all income status of the Processing-Distribution Aggregate for the years 1947 and 1954 appears to have been somewhere between those of Farm Supplies and Farming, as indicated in the tabulation that follows:[27]

	1947 (Billions)	*1954* (Billions)	*% Change*
Net sales	$79.1	$109.1	37.9%
Costs of goods and services	74.3	105.6	42.1
Net profit from operations	4.8	3.5	—27.1
Other income or deductions	0.1	0.1	0.0
Net profit before federal income taxes	4.9	3.6	—26.5
Provisions for federal income taxes	2.0	1.9	— 5.0
Net profit after taxes	2.9	1.7	—41.4
Net profit after taxes as % of net sales	3.7%	1.6%	—55.6%

[26] Computed on the basis of an average hourly industrial wage of $1.24 and $1.81 for 1947 and 1954, respectively.

[27] Derived from Table A-2, Appendix II.

The net sales of Processing-Distribution increased by $30 billion, or 37.9%, from 1947 to 1954.[28] Even so, net profits from operations declined from $4.8 to $3.5 billion, a decrease of 27.1%, and net profits after taxes from $2.9 to $1.7 billion or 41.4%. The result of the combination of increased net sales and decreased net profits after taxes was a 55.6% decrease in the net profit-net sales ratio.

As one might expect, the tendency of net income to lag behind the growth of sales in the Processing-Distribution Aggregate is reflected in the wage policy of this unit. Because the profitability of this aggregate in general is less than other manufacturing sectors of the economy, the Processing-Distribution Aggregate is financially less able to maintain wage rates at the level of most of the other manufacturing sectors. The average hourly earnings figure of workers in the principal component industries of this aggregate was below that of the average for All U.S. Manufacturing, both in 1947 and in 1954 (see the following tabulation).[29] In general, the food industry was paying higher wages than those relating to fibers and tobacco.

	Average Hourly Earnings		
	1947	*1954*	*% Change*
Food and Kindred Products	$1.12	$1.67	49.1%
Tobacco Manufactures	.90	1.30	44.4
Textile Mill Products	1.04	1.36	30.8
Apparel	1.13	1.35	19.5
Leather and Leather Products	1.05	1.38	31.4
All U.S. Manufacturing	1.24	1.81	46.0

Having examined the nature of agribusiness in quantitative terms, we shall now consider it in qualitative terms.

Qualitative Considerations

Organizational Structure

Brief reference already has been made to the decentralization and segmentation of agribusiness in terms of organizational structure. In this respect, too, there is variation as between aggregates. Of course, the Farming Aggregate has the largest number of units—approximately 5 million according to census determination. About 60% of these, however, are so small as to produce little for the market—only about 10% of all agricultural products marketed all told. At the other extreme are a very few large farms with assets in excess of a million dollars. In 1949 there were 100,000 farms with gross operating income in excess of $25,000. The average net income of the group was about $10,000. Together these farms produced about 25% of the total farm products marketed.[30]

Another characteristic of the Farming Aggregate is its limited amount of vertical integration. Of course, one might expect this to follow from the large number and small size of its producing units. Such vertical integration as has taken place has been accomplished in large measure through farmer-owned cooperatives. Approximately 20% of all farm commodities and all purchased farm supplies are handled cooperatively at one stage or another in the production-distribution cycle. However, if one takes into account all operations involved in the production and handling of farm supplies and in the processing, storage, and distribution of food and fiber products, only about 5% would appear to have been done through a mechanism which is vertically integrated with farming.

Basically farming in the United States has been and still is a family enterprise, and a strong sentiment prevails for continuing this pattern. In general, well-equipped family farm units have demonstrated a great capacity to utilize technology and continue in efficient competition with large units.

In contrast, the Farm Supplies and Processing-Distribution Aggregates have integrated much more extensively by expanding into functions closely related to initial activities. A recent study indicates that the Food and Kindred Products industry has entered into more mergers than most other industry sectors.[31] Along with this integration and diversification has come a concentration of a larger proportion of agribusiness functions in fewer firms. In 1955, 76 food and fiber companies

[28] These totals differ from those shown in the Agribusiness Flow Chart in that they reflect transactions rather than consumer purchases.

[29] See *The Economic Almanac*, pp. 243 and 248–253.

[30] See Table A-3, Appendix II.

[31] *Business Conditions*, July 1955, published monthly by the Chicago Federal Reserve Bank.

were listed among the 500 largest American corporations.[32] In 1947, 4 meat packing firms did 41.3% of U.S. total meat slaughtering; 8 flour milling firms, 40.6% of total flour grinding; and 8 cotton fabric firms, 22.2% of cotton fabric merchandising.[33] In 1955, 4% of the food retailing stores accounted for 43.5% of food retailing sales.[34]

The farm supplies companies also have evidenced a tendency toward increased size. In 1955, 13 farm supplies companies were listed among the 500 largest American corporations.[35] In the latest completed Census of Manufactures (1947),[36] 8 companies produced 26.7% of all prepared animal feeds. In 1955, 8 farm machinery and equipment producers accounted for roughtly 70% of the total dollar volume of the industry.[37]

Accompanying these developments has been a marked expansion toward factory preparation of foods in order to reduce the work of the housewife in the kitchen. Also, improved refrigeration and other methods of food preservation have made it possible for specialty foods and seasonal products to be made available throughout the country and around the calendar.

The trends toward bigness and integration in the Farm Supplies and Processing-Distribution Aggregates are presented here as fact with no attempt to prejudge their merits. Before such judgment can be made properly, there is need for an objective appraisal of these trends within an agribusiness frame of reference and against the background of our national economic goals. Such a study should seek to ascertain not only whether such trends have negative results in terms of national economic objectives, but also whether they are capable of providing greater positive benefits in terms of enhancing basic economic stability within the Farming Aggregate, improving human nutritional standards, expanding foreign trade, and the like. (This subject will be discussed further in Chapter 4.)

[32] *Fortune*, July 1956. Supplement.

[33] The tabulation of the 1954 Census of Manufactures was not complete at this writing so 1947 figures were used. *Statistical Abstract, 1955*, p. 802.

[34] "Facts in Grocery Distribution," *Progressive Grocer*, 1956, p. 12.

[35] *Fortune*, July 1956. Supplement.

[36] *Statistical Abstract, 1955*, p. 802.

[37] U.S.D.A.

The Strength of Agribusiness

Without doubt the most far-reaching benefit accruing from the evolution of agribusiness has been the release of workers from agriculture for employment in other occupations, a development made possible by the increased efficiency in the production and marketing of food and fiber. Because food and clothing are primary among human needs, it is essential that man first allocate to the production of these items such time and effort as may be required, leaving for other pursuits the residue of his labor force. Thus, increased efficiency in agriculture and the release of manpower for other work are first prerequisites for industrialization. Moreover, the rate of progress toward increased farm productivity serves to set a ceiling on the rate of industrial development by governing the total manpower available for off-farm work. Fortunately for us, this ceiling has advanced progressively, decade by decade, until today 88% of our working force is available for off-farm jobs and 65% for work entirely outside agribusiness. Also fortunate has been the corresponding ingenuity, courage, and success of industry in venturing into new productive enterprises that provide jobs for the released force.

These twin products of research and technology —the release of farm manpower and the corresponding creation of new off-farm jobs—have provided the basis for our tremendous economic growth and development during the past century and a half. Of course, the key to such growth and development is increased productivity per worker per day, which provides the basis for the creation of new products and wealth. This, in turn, becomes translated into new risk capital, new factories, new jobs, increased consumer purchasing power—which together provide the basis for further economic progress.

The trend toward agribusiness has contributed on both sides of this development by releasing farm labor and generating new off-farm jobs in the industries created to manufacture farm supplies and process and distribute food and fiber products. An important product of this progress is the gradual elevation of the standard of living of all economic groups—both on-farm and off-farm. This improvement in living standards has many attributes. In the case of food, the public has been offered improved quality, greater variety and selection, greater food values in terms of the quantity that can be purchased with an average hour's wages, and less drudgery in the preparation of food in the home. Technology also has increased the mobility of people by freeing them from their own local food supplies; eliminated the fear of famine from the American continent; and given us promise of being able to feed well, for decades to come, a rapidly growing population without added land resources. In the case of fibers, agribusiness technology has reduced the general price of clothing in terms of what an hour's wage will purchase, increased the variety of fabrics from natural and synthetic sources, made possible popular-priced ready-made clothing, and provided fabrics for numerous industrial uses. In addition, living standards for rural people have been further improved by the reduction of back-bending labor and by an increase in time available for education and recreation.

Also, from agribusiness progress has come a strengthened security position for the United States in the event of war or the threat of war. This exists in the form of a reserve capacity to expand food and fiber output to meet emergency needs and in the ability of agriculture to continue to shrink its labor force, even during wartime, thus making added manpower available to the armed forces and war industries. Such contributions proved invaluable during both World War I and World War II. Paralleling this improved security position has been a steadying of the hand of the United States in foreign affairs, particularly during periods of food shortages in other regions of the world.

These benefits resulting from agribusiness progress serve to show that such progress has contributed immeasurably to the general well-being of the American people. Without it we still would be an agricultural nation, unable to devote more than a fraction of our manpower to other than satisfying the primary needs for the maintenance of human life.

Imperfect Evolution

While the evolution of agriculture to agribusiness has been dynamic, bringing with it inestimable benefits to mankind, it also has been imperfect, leaving in its wake serious eddies of stagnation and problems of maladjustment. As pointed out in the introduction, these evolutionary imbalances and imperfections, such as glutted markets, unstable prices, uneconomic farm units, poor managerial training, and lack of agribusiness policy and research formularization machinery, tend to converge about two poles, (1) that of commercial agriculture and (2) that of low-income farmers. While, essentially, both sets of problems are the outcome of the same technological forces applied to agribusiness, the former focuses inward to the use of resources by persons remaining on the farm, whereas the latter looks outward in terms of adjustments in the lives of families and individuals who must seek part-time or full-time employment off the farm if they are to find opportunity for the realization of their productive potential.

Probably the most crucial problem—or more accurately stated, complex of problems—facing commercial agriculture is that commonly referred to as the cost-price squeeze. This phenomenon not only conditions the income status of farm families, but also affects the rate of capital formation and hence the rate of progress in the Farming Aggregate. Basically, this situation is the result of the inability of farm management, itself, to tailor output to market demand in the same manner as does the management of business and industrial enterprises.

Commercial Agriculture. A careful look at our farm economy in an agribusiness setting will reveal that it is one segment in which it is difficult for management, by virtue of its own action to relate output accurately to market demand, at a price compatible with national economic price levels. Of course, all industries tend at times to be subjected to temporary market gluts such as the build-up of automobile stocks in 1955 and 1956; but if such continue, or give promise of continuing, management itself cuts back the rate of output until a balance between supply and demand is restored at a price level in line with national economic trends. But in agribusiness, in the event of a market glut, such restoration of balance does not take place promptly as the result of the decision of management. Instead, frequently during peacetime periods surpluses of raw products continue to pour off our farms year after year, even in the face of falling prices.

This seemingly irrational behavior of the food and fiber segment of the economy is the result of several factors, some inherent in the nature of farm production and some the result of the organizational structure of agribusiness. The fact that agricultural output is tied closely to weather and seasons makes it impossible for farm management to gear production to a goal with the same precision as does industry in general. Also, the fact that farming can be and is done efficiently on relatively small units means that it is characterized by several million production units, each of which is too small to integrate supply and demand functions. In addition, because fixed farm investment is high, farmers tend to produce as long as they can cover their out-of-pocket costs, including essential family living items. Also, research and extension education often have stimulated greater output in the face of market gluts by their relatively stronger emphasis on efficiency and low unit costs than on market expansion.

Further contributing to this exceptional behavior of agribusiness with respect to market gluts is the segmented structure of the food and fiber economy. Particularly isolated are many of the on-farm decisions with respect to those made off-farm. To illustrate, because today off-farm businesses manufacture and sell to agriculture more than half of all farm inputs, such businesses are a decisive factor in determining farm production, year by year. By constantly improving the quality of farm supplies and stepping up selling methods to promote their use, business adds to the productive capacity of agriculture. In large measure the decision by business to pursue such policies is based on considerations internal within the supply firms rather than upon the needs of the food and fiber markets for increased production. Nor are proprietary businesses alone in this type of behavior, for farmer-owned cooperative suppliers frequently do likewise. For example, a farmer cooperative feed company may attempt to help its members cut feed costs by waging an aggressive sales campaign without simultaneously helping them to analyze the resulting effect of increased output on the price of animal products and net farm income.

On the marketing side of agribusiness, business firms handle virtually the entire output of agriculture in terms of processing and distribution. In general, these firms stand ready to receive whatever volume agriculture turns out. However, in doing this they are guided largely by considerations internal to the firms involved, rather than out of consideration for the needs of agriculture for expanded outlets. Not only is the agribusiness structure segmented with reference to on-farm and off-farm operations, but also with respect to certain off-farm functions. This particularly is true in the processing-distribution phase where one firm may assemble a farm commodity into sizable stocks, another may provide storage, another may process, and another may serve as distributing agent. At each step the logical and customary policy is for the firms involved to make decisions on the basis of internal considerations rather than with reference to the needs of agriculture or the public.

The result is that neither the farm nor nonfarm phase of agribusiness has developed counter forces to offset the inherent weaknesses of farming—weaknesses rising out of the inability of agricultural management to relate supply to demand in a manner comparable to that existing in industry. In fact, sometimes the behavior of the nonfarm segments of agribusiness may even tend to have the opposite effect, as for example when a grain firm withholds its storage capacity at harvest time by staying out of the market until prices drop. On the other hand, many handlers, particularly farmer-owned organizations and firms that process, do make an effort to stabilize the market by buying more heavily than for current needs during periods of market gluts in an effort to stabilize commodity prices or to protect their own inventory position, or both.

These facts are presented, not to show that one segment of agribusiness is more self-centered than another, but to point out the effect of the segmented nature of the agribusiness decision-making process in contrast to that of an industry which is highly integrated, vertically. The point is that agribusiness has not developed within itself the counter forces essential to offset the inherent weakness of the Farming Aggregate with respect to relating supply to demand when faced with a market glut.

The adverse consequences of this inherent inability of agriculture to adjust supply to demand readily have gained increased significance in recent years as the result of the trend toward relatively more and more purchased inputs. The reason for this is twofold: the tendency of prices of purchased supplies to follow a relatively inflexible course, even when farm prices decline markedly,[38] and the tendency of such supplies to add further to the output capacity of the farm plant. These factors, plus the inherent inelastic demand[39] for farm commodities as a whole, mean that agriculture frequently is confronted with market gluts which cause commodity prices to decline irrespective of farm costs. Such is the basis of the cost-price squeeze. While in the long run such maladjustments do tend to be resolved by economic forces under a free market system, the inherent greater weakness of agriculture relative to industry when it comes to action by management, itself, means that such adjustments come slowly and with much suffering on the part of farm operators.

[38] The behavior of supply prices is explained in large measure by the nature of costs within the industry, plus the ability of management to tailor output to demand.

[39] Kenneth Ewart Boulding, *Economic Analysis* (New York, Harper and Brothers, 3d edition, 1955), p. 778, comments on the principle as follows:

"Unprofitability of Industries Faced with Inelastic Demands

"Some interesting corollaries follow from these propositions. The first is that in a progressive society industries whose commodities suffer from an inelastic demand always tend to be relatively unprofitable, and technical progress in these industries if anything accentuates this unprofitability. In a progressive society the proportion of resources employed in such industries continually undergoes a relative decline. The way in which society brings about such a relative decline, however, is by making the unfortunate industry relatively unprofitable and so squeezing resources out of it. The case of agriculture is an important example of this tendency. Agriculture is chronically unprofitable in a progressive society. Even though the past two centuries have seen an enormous expansion of agriculture, the increase in other industries has been much more marked; the very fact of the 'drift to the towns' proves that agriculture has been less profitable, broadly, than industrial occupations. Moreover, technical improvements in agriculture, although they benefit those who introduce them first, do not ultimately have the effect of making agriculture more profitable. They may even have the opposite effect, for by hastening the rate of progress they bring about a still greater pressure to force resources out of agriculture. This pressure is exercised through the price system; technical progress in agriculture brings about an increase in production. Then because of the inelastic demand for agricultural products, their prices are forced down until many of those who were previously engaged in agriculture are forced out of it. The forcing-out process may take a long time, and meanwhile the industry may be quite generally unprofitable, even for the innovators. Moreover, by the time equilibrium would have been reached and agriculture restored to normal profitability again, new technical improvements may start the process all over. In eras of rapid technical change, therefore, agriculture may be continuously unprofitable."

Low-Income Farm Families. As already indicated, the solution to the problem of low-income farmers is one of progress in a different direction from that of commercial agriculture. Approximately 60% of the 5 million farm families listed by the census fall more or less into this category. Of these 3 million families, 1.5 million received under $1,000 cash income in 1950.[40] Thus, the problem of low-income families in agriculture is a sizable and serious one. Not only does permitting this situation to go unsolved subject these persons to a substandard level of living, but society loses the value of their productive potential. Viewed in broad perspective, this problem also is the result of a lag in the adjustment from the subsistence farming era to the agribusiness farming era. By the same token, the answer is to assist these families more quickly to adapt themselves to the new era. Because of the trend toward fewer and larger agricultural units, there is only limited opportunity for low-income families to become efficient and profitable farm operators. While a certain number of able, younger farmers can solve their problem by "climbing the agricultural ladder,"[41] these will become more and more the exception to the rule. The majority of the low-income families must seek to increase their earnings by part-time or full-time productive work off-farm. Since such jobs must be provided by business, the low-income problem in agriculture, also, is an agribusiness problem.

As will be elaborated in Chapter 4, a solution must be worked out jointly by various groups, including responsible representatives of farm organizations, business firms, civic groups, extension service, vocational educators, and employment services. While this is not the place to debate the merits of rural living vs. city living,[42] it can

[40] *Development of Agriculture's Human Resources*, U.S.D.A., April 1955, p. 1.

[41] There is a capitalization problem for both the commercial and the low-income farmer, namely, how to handle inheritance problems such as that of recapitalization by every generation. As land, livestock, and machinery investment rises, the need correspondingly increases for research to find a new approach to provide greater continuity of the capital structure of the Farming Aggregate.

[42] Many rural leaders, and others, have considered the farm home an ideal place to rear a family. They have argued that the decrease in farm units tends to weaken the social and moral fiber of the nation. Some even have contended that low-income farmers should be retained in agriculture, by subsidization if necessary, in order to preserve a rural heritage to a sizable portion of the population.

be pointed out that the transfer of low-income farmers to industry need not deny these or other families the privilege of living in a rural or semi-rural community. With modern transportation and power equipment for yard maintenance, gardening, etc., families can combine urban employment with rural living. Actually, with the higher earnings from more and more productive employment, the living standard will tend to be considerably higher on such a basis than on unproductive small farms.

The Role of Government. When the so-called farm problems are viewed with agribusiness perspective, it becomes clear that the trend toward governmental assistance to agriculture is the result of inherent weaknesses in the food and fiber economy, rather than merely the consequence of the efforts of socialistic promoters. This is true with respect to the problems both of commercial agriculture and of low-income farmers.

The cost-price squeeze is a problem of long standing. It was behind the Granger Movement of the 1870's and the Populist and Free Silver Movements of the 1890's. Following World War I, it first was reflected in a drive for cooperative marketing and the pressure for the McNary-Haugen type of legislation. A little later the mounting pressures culminated in passage of the Agricultural Marketing Act of 1929, which created the Farm Board, and the Agricultural Adjustment Acts of 1933 and 1938. Following World War II, the pressure for remedial legislation has continued unabated with the result that most previous programs have been expanded and new ones added such as the acceptance of foreign credits in exchange for exports and the soil bank. The magnitude of the drive in this direction is evidenced by the fact that currently government is assuming a larger and larger role of this type, despite a conscious effort by the Eisenhower Administration to move in the opposite direction. This trend is not likely to be reversed as long as agriculture continues to have an excess capacity to produce and as long as the agribusiness segment of the economy lacks the inherent ability within itself to equate supply and demand through decisions made by management.

Nor do pressures for governmental assistance come only from farm people. In time government programs tend to become part of the organizational and operating scheme of all phases of agribusiness, including the Farm Supplies and Processing-Distribution Aggregates. For example, those engaged in commodity storage have almost as much a vested interest in government programs as those engaged in farming.

But at least until 1957 governmental aid really has only alleviated problems and not solved them. The production side of agribusiness still continues to expand output faster than the marketing side develops outlets at prices compatible with general economic conditions. The weight of evidence is that in the United States agricultural production will continue to increase at a rate in excess of demand for at least a decade—probably longer. Hence the problem of commodity surpluses is likely to be a continuing one. At least the probability of this surplus problem's continuation is sufficiently great to make it prudent to proceed on this assumption. This subject will be discussed further in Chapter 4.

Research. Research in food and fiber today stands at an all-time high. However, this is segmented in that it does not view specific projects within the over-all agribusiness framework. Also when related to total dollar sales, this research is taking place at a rate considerably below that prevailing in American industry. In general, nonfarm components of agribusiness, such as those processing food, fabricating textiles, and manufacturing farm machinery, spend less on research, relative to dollar sales, than nonagribusiness industries.[43] In spite of large government expenditures for farm research,[44] this part of agribusiness also lags behind the total economy. In fact the Department of Agriculture estimated in 1956 that industry spent on the average $2 per $100 of sales for research whereas agriculture only spent $0.50 per $100 of sales.

Summary. In brief, agribusiness is a new "creature" which has evolved from the past under the impetus of technology to assume the role previously performed by agriculture alone. Basically, it

[43] See Tables A-4, A-5, and A-7, Appendix II.

[44] See Table A-6, Appendix II.

is superior to its "ancestors," bringing with it vast benefits to mankind. However, its progress has been uneven and presents us currently with numerous problems in the form of areas of maladjustment. To overcome these, the United States, like most other countries around the globe, has resorted to governmental assistance of various types. In general, these have alleviated rather than solved the problem.

Having in this and the preceding chapter looked at agribusiness from the *outside*, so to speak, we shall attempt in the next chapter to view it from the *inside* against the background of the national economy by means of *input-output analysis*.

CHAPTER 3

Agribusiness and Input-Output Economics

HAVING described the evolution of agribusiness in Chapter 1 and having discussed its general nature in Chapter 2, the authors in this chapter attempt to analyze, in terms of 1947 data, the inner workings of the food and fiber sector of the economy. The basic methodology is that of tracing the flow of resources, goods, and services among major agribusiness enterprises and relating this flow to the national economy.[1] Such analysis, which depicts and quantifies the actual flow of goods and services through the economy in a given time interval, is in contrast to the more conventional technique of rationalizing or generalizing from scattered or random observations.

This chapter is divided into three sections. The first, *Input-Output Analysis*, describes the methodology and explains the procedures involved in using the Agribusiness and National Economy Matrix (Exhibit 2); the second, *Agribusiness Interactions*, describes, by use of pie charts, tables, and flow charts derived from Exhibit 2, the important components of agribusiness and their direct relationships with one another and with the rest of the economy; and the third, *Direct and Indirect Agribusiness Relationships*, deals with the reciprocal and rebounding effects of agribusiness industries upon one another and upon the total economy.

INPUT-OUTPUT ANALYSIS

As a method of analysis, the authors have utilized the technique of Input-Output Economics as developed by Wassily W. Leontief, Professor of Economics, Harvard University. This method, sometimes referred to as Interindustry Theory, is particularly useful because of the complexity of the interrelationships within agribusiness. In the past, economic theorists have relied heavily on concepts of supply and demand, wages, prices, and interest rates to explain the material operations of our society. Such analyses of man's economic behavior leave largely unexplained the actual interplay of forces in the day-to-day operation of the economy. The Leontief input-output technique serves as a useful supplementary tool to illustrate such actual flows of goods and services, thus supplying an additional dimension to economic analysis. Through cooperative effort in recent years, the Bureau of Labor Statistics, the Bureau of Mines, the Department of Commerce, the Bureau of the Budget, the Department of Agriculture, and others have assembled and codified a vast amount of material suitable for such methods of analysis, particularly for the year 1947.[2]

As in Chapter 2, it has been necessary to rely on value judgments in deciding which industries

[1] The year 1947 was selected for this purpose because it is the most recent census year for which reasonably complete data were available. A similar study has not been tabulated for the census of 1954 as of this writing.

[2] The reader will find descriptions of the input-output method in the following books and articles:

Wassily W. Leontief, *Studies in the Structure of the American Economy* (New York, Oxford University Press, 1953).

——— *The Structure of the American Economy, 1919–1939* (New York, Oxford University Press, 2nd edition, 1951, especially Part IV).

Conference on Research in Income and Wealth, *Input-Output Analysis Technical Supplement* (New York, National Bureau of Economic Research, 1954).

Studies in Income and Wealth, Vol. 18, *Input-Output Analysis: An Appraisal* (New York, National Bureau of Economic Research, 1955).

W. Duane Evans and Marvin Hoffenberg, "The Interindustry Relations Study for 1947," *The Review of Economics and Statistics*, May 1952.

Earl O. Heady and G. A. Peterson, *Application of Input-Output Analysis to Agriculture* (Iowa State College, Research Bulletin 427, April 1955).

Karl A. Fox and H. C. Norcross, "Agriculture and the General Economy," *Agricultural Economics Research*, January 1952.

"Discovering How U.S. Economy Fits Together and Grows," *Business Week*, March 31, 1956.

or parts of industries along the perimeter of agribusiness to include in the analysis. In general, the matter has been resolved by allocating parts of industries to agribusiness on the basis of the proportion of goods and services that a given industry bought or sold in connection with the production, processing, or distribution of agribusiness goods. (See Exhibit 2.) A detailed analysis, based on input-output data for a more recent year than the 1947 input-output study, would enable one to attain a more accurate picture of how agribusiness actually operates today and some of the changes that take place in these relationships over time. However, because such data are not currently available and because of the dangers inherent in projections, the authors have chosen the alternative of examining agribusiness in terms of 1947 statistics. (See Exhibit 2.)

Input-output Exhibits 2, 13, 14, and 15, derived from the codified material mentioned above, provide a description of relationships among various sectors[3] of the American economy for the year 1947. Such analysis is helpful in providing a better understanding of how our economy operates. Also, these input-output figures provide a basis for estimating the impact of a given change in one sector of the economy upon the flow of goods and services among other sectors. Such prediction, of course, assumes either that the structural relationships among the various sectors remain constant as between the base and projected time periods or that one can successfully anticipate trends and inject corrective factors into the analysis.

Applying this principle to input-output data for the year 1947, one should be aware of such limitations as the following:

(1) The year 1947 was not representative or "normal" with respect to certain crops, industries, and relationships within agribusiness, and no single year is likely to be.

(2) Direct linear relationships do not exist for certain industry sectors. Nor does a single year provide an adequate basis for determining corrective factors for use in projecting trends. Therefore, while the available data for the year 1947 are useful in developing the nature of agribusiness relationships, they are inadequate for predicting future trends.

(3) The aggregate process used in this study for summarized sector data is so general as to obscure many of the peculiar characteristics of individual industries within sectors and their relative importance to other sectors.

Explanatory Notes for the Agribusiness and National Economy Input-Output Chart (Exhibit 2)

Exhibit 2 depicts, in summary form, the dollar value flow of resources, goods, and services through agribusiness and the rest of the economy in 1947, both by industry of origin and by industry of destination. The horizontal rows, as read from left to right, trace how the output of each sector of the economy is distributed among other sectors. The vertical columns, as read from top to bottom, trace how each industry obtains or purchases its needed inputs of goods and services from the other sectors.[4]

In an ideal matrix or input-output chart, each industry sector would appear in identical form, both as a selling entity on the side and as a purchasing entity at the top of the chart. The totals of each sector—if expressed in dollar terms—would be equal when added vertically and horizontally, and the sum of all sectors would be the input and output totals for the national economy. However, a matrix in such detail would tend to be unwieldy for the purpose of this study in that it gives unnecessary emphasis to nonagribusiness phases of the economy. Hence, the authors have designated in Exhibit 2 only those industry sectors and aggregates that are of prime importance to agribusiness—grouping other transactions into summary sectors. Where sectors appear at both the side and top of the figure and have the same title and input-output code number, they are identical and contain the same aggregations of industries

[3] See Appendix I for definition of sector and other terms used in the input-output charts. An input-output chart may be likened to a double-entry bookkeeping system which shows purchases from and sales to each of the sectors of the economy. (Exhibits 13, 14, and 15 are bound in the back of the book.)

[4] This manner of tracing industry's inputs and outputs is the origin of the "input-output" chart or table. The data for these sectors represent aggregations from data prepared in much greater detail by the Bureau of Labor Statistics under the direction of Professor Leontief. The authors, assisted by Herman Southworth, prepared this agribusiness chart and selected and aggregated certain industries that together, in our opinion, comprise the main components of agribusiness and show their relation to the rest of the economy.

(i.e., the Farming, Food Processing, and Fiber Processing Aggregates).[5]

On the other hand, the inclusion of a sector only on the side or top of the chart, but not on both, means that it was significant to agribusiness only as a buyer or a seller, and not as both. From this it follows that industries which appear as sectors and aggregates (rows) only on the selling side (e.g., Containers, Fertilizers, Power, etc.) are included in the All Other Industries Aggregate (column) on the purchasing side. Similarly, industries which appear as sectors (columns) only on the purchasing side (e.g., Eating and Drinking Places, Soap and Related Products) are included in the All Other Industries Aggregate (row) on the selling side. Thus, while in essence the whole economy is represented in the flow of goods and services shown in this matrix, this specific arrangement of sectors more or less limits the use of such matrix to the particular purposes of this study.

To facilitate an understanding of Exhibit 2, let us examine the sector Food Grains and Feed Crops, row 4,[6] and trace selected input-output relationships pertaining to it. Reading from left to right along row 4, the transactions reflected in each column may be viewed either as sales or purchases (depending on whether they are approached from the viewpoint of the sector shown at the left or top of the table). To illustrate, column 1 of row 4 shows transactions between the Food Grains and Feed Crops sector and the Meat Animals and Products Industry sector in the amount of $3,841 million. From the standpoint of the sector shown on the left, this was a sale, but from that of the sector shown at the top, it was a purchase. In the explanation that follows, we first shall look upon transactions from the former viewpoint, thus considering them as sales.

The line of figures at both the side and top of Exhibit 2 indicates the sequence of sectors as arranged in that matrix, whereas the figures in parentheses following each section denote the code used in the Interindustry Relations Study of 1947, referred to earlier. Proceeding from left to right along row 4, the Food Grains and Feed Crops sector of agribusiness sold $1,279 million of output to the Poultry and Eggs sector (column 2); and $1,755 million to the Farm Dairy Products sector (column 3). Column 4 indicates that the Food Grains and Feed Crops sector utilized its own products in the amount of $817 million (chiefly seed). Continuing along row 4, the Food Grains and Feed Crops sector sold $33 million of output to the Cotton sector (column 5); $20 million to the Tobacco sector (column 6); $33 million to the Oil-Bearing Crops sector (column 7); $85 million to the Vegetables and Fruits sector (column 8); and $63 million to the All Other Agriculture sector (column 9). Similarly, column 10 shows that zero volume of Food Grains and Feed Crops was utilized by the Fishing, Hunting, and Trapping sector.[7] Column 11, as indicated by its title, Farming Aggregate, presents the sum of all values shown in columns 1 through 10.[8] Columns 12 through 38 show sales by the Food Grains and Feed Crops sector to other industry sectors or aggregates. The All Other Industries Aggregate (column 39)[9] is comprised of those industries which purchase only minor amounts of goods and services from agribusiness sources.

The first 39 columns, together, represent all the processing or intermediate sectors of the economy, whereas the next 5 columns represent categories of end-product demand.[10] This end-product demand represents the final consumption of the goods and services produced in the processing sector. For example, the Food Grains and Feed Crops sector sold $919 million to the Exports (minus Competitive Imports) sector (column 40);[11] less than $0.5

[5] The Farm Supplies Aggregate is identified only on the sales side since, in the original study, many items such as farm machinery were included with depreciation in the Labor and Capital Utilization sector. See footnote 20, p. 28.

[6] This number, as do all the numbers in parentheses of the particular industry sectors in Exhibit 2, refers to input-output categories of particular industries defined for use in all input-output studies. See Appendix I for more detailed descriptions of each industry sector and aggregate used in Exhibit 2.

[7] As explained in Chapter 2, the Farming Aggregate as a component part of agribusiness does not include the Fishing, Hunting, and Trapping sector.

[8] As indicated earlier, the selection of sectors and aggregates and the grouping of functions within them have been done arbitrarily by the authors for the purpose of facilitating analysis. While this arrangement is adapted to our general purpose, other arrangements might prove more appropriate for other types of studies.

[9] See Appendix I for industries included in the All Other Industries Aggregate.

[10] See page 50 for additional explanation of end-product sectors.

[11] The Exports (minus Competitive Imports) sector (column 40) includes all exports and such items as foreign purchases of U.S. ocean transportation, air transportation, motion picture royalties, insurance payments for foreigners, income on investments abroad, etc.; minus all imports (at the foreign port value) and a number of other items such as cost to the United States of foreign ocean transportation, travel abroad, off-shore purchases by the Federal Government, gifts, and other unilateral payments.

million to the Government Purchases sector (column 41);[12] and zero quantity to the Gross Private Capital Formation sector (column 42).[13] The Food Grains and Feed Crops sector also sold $45 million of goods and services directly to consumers as shown in the Consumer Purchases sector (column 44).[14] The final vertical column (45) is entitled Gross Domestic Output and represents the value, at producers' prices, of all the output for each industry sector.[15] Food Grains and Feed Crops had a gross output of $11,003 million.[16]

Having examined the output of the Food Grains and Feed Crops sector as distributed among the various processing and end-product sectors of agribusiness and the national economy in 1947, we shall next consider the vertical column of this sector and trace through the purchases.[17] Continuing the use of the Food Grains and Feed Crops sector as an example (following column 4 from top to bottom) one will note that this sector purchased no goods and services from the Meat Animals and Products sector (row 1); the Poultry and Eggs sector (row 2); or the Farm Dairy Products sector (row 3). Column 4, row 4 relates to the same transfer of goods within the Food Grains and Feed Crops sector mentioned earlier; however, when read down the column, the transfer shows up as a purchase of $817 million of supplies rather than as a sale. Similarly, following down column 4 from rows 5 through 54, the purchases (inputs) by the Food Grains and Feed Crops sector from the respective sectors of the processing or intermediate industries may be ascertained. Row 55, All Other Industries Aggregate, includes those industries supplying only a small amount of goods and services to the principal agribusiness industries.[18]

The remaining three rows (56–58) represent factor payments made by the purchasing columns. Factor payments consist of cost items not included in the transactions of the processing sectors. For example, the Food Grains and Feed Crops sector made no purchases of Noncompetitive Imports (row 56). However, Food Grains and Feed Crops did pay $348 million in taxes (Government Services sector, row 57)[19] and $6,273 million to the Labor and Capital Utilization sector (row 58) in the form of wages, salaries, interest, management services, and depreciation, of which $853 million was labor cost.[20]

[12] The Government Purchases sector shows purchases by the government from the processing sectors for services, wages, salaries, supplies, and equipment.

[13] The Gross Private Capital Formation sector includes the total value of all new private construction and producers' durables. In the case of agribusiness, such durables would include tractors, combines, implements, etc. In essence, the Gross Private Capital Formation sector is an attempt to establish a capital expenditure sector within the final demand part of the interindustry chart. Hence the Farm Machinery Industry sector was not separated out in Exhibit 2 as its output was placed in Gross Private Capital Formation rather than distributed to the respective Farming sectors.

[14] The Consumer Purchases sector (subtitled "Household" to conform with the nomenclature used by Leontief in other input-output charts) refers to direct consumer purchases of the various goods and services offered by each of the sectors. These figures are in producers' values, wholesale and retail trade were separated out and margins were estimated for each industry sector between producers' and purchasers' values. Therefore, these margins are not included in the Consumer Purchases column as identifiable to any one industry sector but occur as separate sectors. Column 44 is discussed more fully in Appendix III, Section A.

[15] Gross Domestic Output includes receipts for: (a) commodities shipped by business units classified within a particular industry—both primary and secondary products; (b) primary products of the industry; (c) scrap sales, contract and commission work, and electric energy sales; (d) competitive imports at domestic port value; (e) additions to inventories of finished products.

[16] All the entries in our input-output figures are in producers' (rather than purchasers') values. Gross output, as illustrated here, is the value before addition of any marketing charges for Food Grains and Feed Crops.

[17] One of the most important final demand sectors is that of consumer purchases because it represents direct personal consumption and without additional processing. Statistically it poses many problems a description of which is found in Appendix III, Section A. The values for each industry sector are in producers' values, and the wholesale and retail margins on the goods consumers buy are treated as separate sectors in the Consumer Purchases column.

[18] See Appendix 1 for industry sectors included in the "All Other Industry" row.

[19] The Government Services sector includes payments to federal, state, and local governments for such items as taxes, Social Security, import duties, postage, etc. The purchases from government-owned plants are included in the appropriate industry sector.

[20] The Labor and Capital Utilization (or Leontief's "Household") sector, as a row, includes wages, salaries, interest, donations, management services, and depreciation. The authors regret that the data from which our figures were derived did not segregate labor cost items from capital items; however, in the original census data such separation was attempted and the labor cost for this particular sector, Food Grains and Feed Crops, was estimated at $853 million. Any future basic input-output studies should separate capital and related expenditures from labor costs as a basis for a more accurate description of the workings of our economy. The authors, using unpublished B.L.S. data, have derived the following estimates to indicate the relative proportion of labor and capital inputs made by the industry sectors comprising the Farming Aggregate. See Table A-36, Appendix II. In addition, one must note that the Farm Machinery Industry sector would sell its output to the Gross Private Capital Formation (column 42) directly, and such purchases would not be designated among the Farming sectors; therefore the Farm Machinery Industry sector is included in the All Other Industries Aggregate (row 55). Also depreciation on such farm machinery would be included in the Labor and Capital Utilization row under each Farming sector column.

Reading down column 4, the final row, 59, shows the total outlays of this purchasing column, covering all costs consistent with gross output. For example, the total outlay of the Food Grains and Feed Crops sector was $11,006 million. The total gross outputs and outlays (inputs) of the processing sectors are equal in an accounting sense, because current outlays, with allowance for profits and inventory changes, are equal to current receipts. Thus, the totals of the Gross Domestic Output (column 45) and the Gross Domestic Outlay (row 59) of the Food Grains and Feed Crops sector were $11,003 million and $11,006 million respectively. (The difference of $3 million is a result of the rounding procedure.) While the totals of the specific processing sectors on the output (rows) side are not identical with those on the input (columns) side of Exhibit 2, a balance is obtained if one compares the sum of all such sectors, including in both input and output totals the aggregate labeled "All Other Industries," which is a residual category including all processing operations not contained in the specific sectors. Likewise, although the individual end-product sectors (columns 40–44) do not match their counterparts in the factor payments sectors (rows 56–58), yet the sum of all sectors comprising the two categories are equal.

With minor adjustments, for statistical and computational differences, the gross national product may be derived from the end-product (columns 40–44) or factor payment sectors (rows 56–58) of Exhibit 2—factor payments being equal to the sum of consumer expenditures, changes in inventory, net investment, net foreign balance, and government expenditure.

AGRIBUSINESS INTERACTIONS

In this section we shall apply to agribusiness the methodology outlined in the preceding section for the purpose of tracing and quantifying the major streams of goods and services through the food and fiber segment of the economy. The first part of this section indicates the manner by which the Agribusiness Flow Chart for 1947 (Exhibit 3) was derived from the Input-Output Chart (Exhibit 2); the second discusses the interrelationships of agribusiness sectors and aggregates; and the third analyzes the private and public consumption of processed agribusiness products and the labor and capital requirements for producing them.

In doing this, we shall attempt to analyze further the basic data of Exhibit 2 by breaking them down and summarizing them in flow charts, tables, and pie charts for the purpose of presenting graphically the movement of goods and services through the agribusiness economy, in the conversion of farm commodities into consumer items. These charts and tables are products of input-output analysis in that they are derived from Exhibit 2. As used here, they are complementary to the basic Input-Output Chart—serving as methods of portraying facts derived by interindustry analysis.

The Agribusiness Flow Chart for 1947

The Agribusiness Flow Chart for 1947, Exhibit 3, illustrates in graphic form much of the agribusiness data contained in the Input-Output Chart, Exhibit 2. In the Flow Chart one can trace the major farm supplies items as they were utilized in farm production, and the movement of the resultant farm commodities through successive stages of processing and distribution. The pattern of this specific flow chart is but one of many that might be devised from basic matrix data to highlight different features of agribusiness or to amplify a specific segment of it.[21]

[21] Certain apparent differences in quantities shown in Exhibits 2 and 3 exist which require a reconciliatory explanation. Probably the most obvious difference is in the breakdown of manufactured farm supplies—the Flow Chart showing this function in greater detail. This results because the basic interindustry matrix for 1947, from which Exhibit 2 was taken, grouped most farm supply items in the general sector labeled "Labor and Capital Utilization." Hence, in Exhibit 2, this same grouping has been retained. However, in the Flow Chart the authors have drawn on U.S.D.A. data for the purpose of presenting a more complete breakdown. Thus, while the same dollar value of farm supplies is included in both charts, the categories in which they appear are different. Since this is the only instance in which the authors have gone outside Exhibit 2 for data in preparing the Flow Chart, all other discrepancies can be reconciled by adjustments in sector grouping. Reconciliations are found in Appendix III, Section B.

EXHIBIT 3. AGRIBUSINESS FLOW CHART: 1947

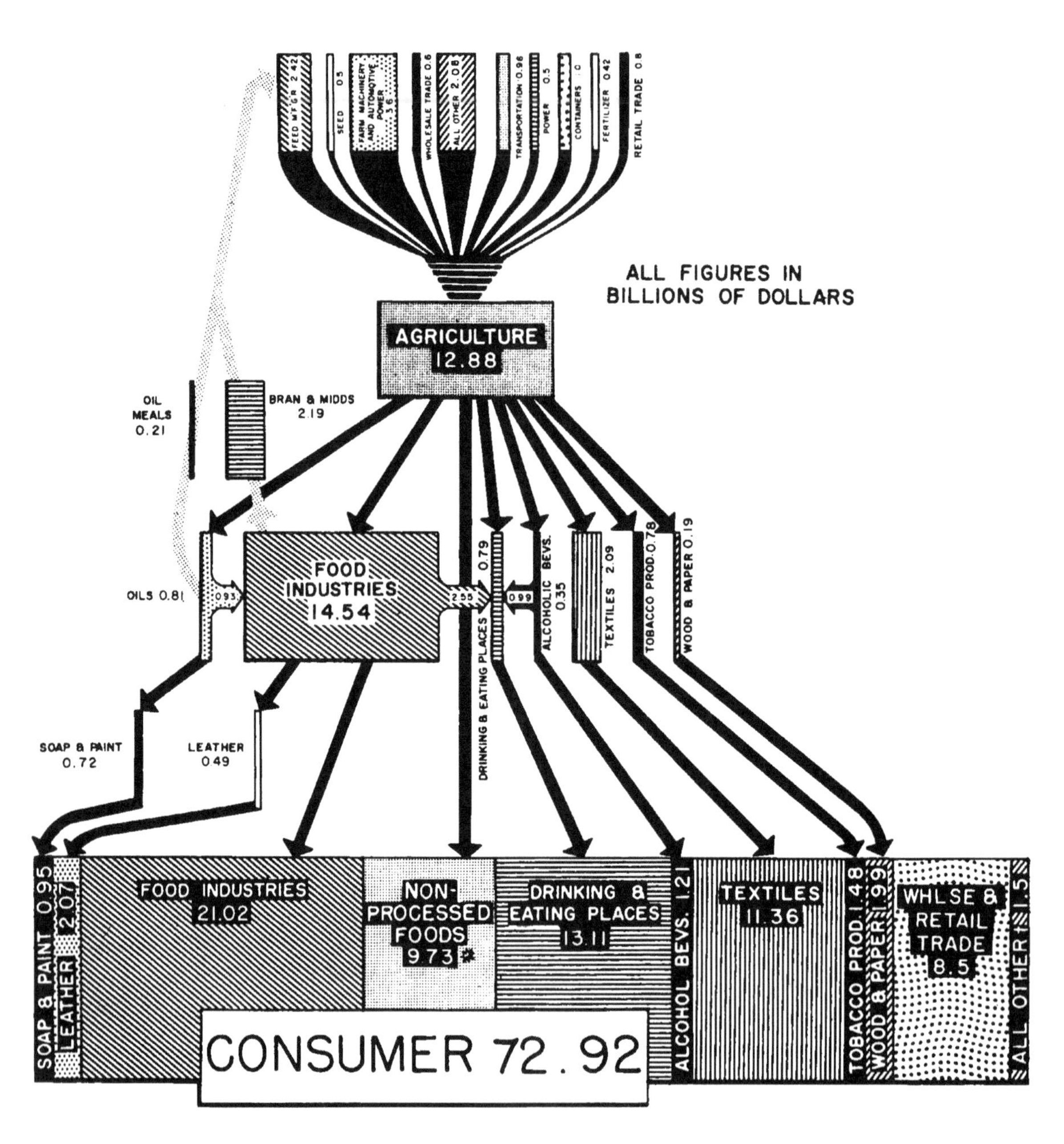

* Includes food products purchased by consumers in relatively unchanged form (like eggs or fresh vegetables) and also those consumed on the farm.

† Includes sea foods, synthetic fibers, and imports.

SOURCE: U.S. Department of Labor, Bureau of Labor Statistics, Division of Interindustry Economics, October 1952.

The data for the Flow Chart, Exhibit 3, and for the Input-Output Chart, Exhibit 2, for the most part, are the same and the source is the 192 Inter-industry Flow of Goods and Services by Industry of Origin and Destination for the United States, 1947. Exhibit 3 is just one of many ways of presenting the data contained in Exhibit 2 to show the interrelationships and intrarelationships that exist within agribusiness and between agribusiness and the rest of the economy. The main purpose of the Flow Chart is to give some general over-all dimensions of agribusiness as it existed in 1947, using the best data and estimates available. This is not a complete or an exact picture of agribusiness, yet it does present a general outline of that part of the economy the authors refer to when using the term "agribusiness."

Sector and Aggregate Interrelationships

Approach

As a means of viewing quantitatively the inner workings and characteristics of agribusiness and its component parts, we have selected the aggregates of *Farming, Food Processing, Fiber Processing*, and *All Other Industries* for further study. Also, we shall examine the sectors comprising them, plus the sector of *Tires, Tubes, and Other Rubber Products*. In addition, we shall consider the factor payment sectors of *Government Services* (Taxes, etc.) and *Labor* and *Capital Utilization* (wages, depreciation, interest, managerial services, etc.) used in agribusiness production and *Government Purchases* and *Consumer Purchases* of agribusiness products.[22] In selecting Farming, Food Processing, Fiber Processing, and Tires, Tubes, and other Rubber Products for more detailed study, two criteria were used: (1) each represented an important phase of agribusiness and (2) each appeared in identical form on both the input and output sides of the matrix (Exhibit 2).

[22] Government Purchases does not equal Government Services because of an unbalanced cash budget. Consumer Purchases is not consistent with Labor and Capital Utilization (but it is used as a "household" column and row respectively in the Leontief nomenclature). Even though these two end-product sectors do not balance the two factor payment sectors, they do provide the reader with an opportunity to view their respective relations to each other and to the agribusiness-oriented sectors. For a more complete description of Consumer Purchases, see Appendix III, Section A.

The All Other Industries Aggregate is included in our analysis to indicate the relative importance to agribusiness of industries not individually represented in the Input-Output Chart. The final demand sectors of Government Purchases and Consumer Purchases and their counterpart, factor payment sectors, are presented to provide the reader with an opportunity to view the impact of these important portions of the economy on the agribusiness-oriented sectors.

Obviously, there are other units of agribusiness that are equal in importance—particularly in the Farm Supplies Aggregate—to those selected. Limitations in the basic data, however, prevented a satisfactory detailed analysis of such units as a part of this study.

The same general method is used in analyzing all units. First the sales and purchases of each aggregate and sector are aggregated and summarized in pie charts. Then sales and purchases are tabulated and ranked in order of dollar volume (this essentially is a rearrangement of the data in Exhibit 2).[23] In addition, each sector of the Farming, Food Processing, and Fiber Processing Aggregates is analyzed in detail by placing sales and purchases in parallel relationships rather than in the perpendicular relationship of the input-output chart (Exhibit 2). Finally, the data are totaled by sectors, aggregate by aggregate, and tabulated in terms of a new composite unit to indicate interrelationships in an agribusiness setting. This new composite is called "The Secondary Agribusiness Triaggregate."[24]

The Farming Aggregate

The Farming Aggregate both purchased and sold over $40 billion[25] worth of goods and services in 1947[26] (see Exhibit 4). Its purchases can be divided into three broad categories; the largest is

[23] Other tabular presentations of each industry sector of the input-output chart (Exhibit 2) are located in Appendix II.

[24] The Secondary Agribusiness Triaggregate is composed of the Farming, Food Processing, and Fiber Processing Aggregates. It differs from the Primary Agribusiness Triaggregate which included the Farm Supplies Aggregate and dealt with food and fiber processing as a single unit. See footnote 1, p. 7.

[25] The $40 billion of sales of the Farming Aggregate include $10 billion of sales to its own sectors. This explains why its sales are larger here than in the operating statement in Chapter 2.

[26] Purchases and sales of this aggregate do not balance exactly as the result of rounding. The source of all the data for Exhibit 4 is the Input-Output Chart (Exhibit 2).

EXHIBIT 4. PURCHASES BY AND SALES OF THE FARMING AGGREGATE: 1947

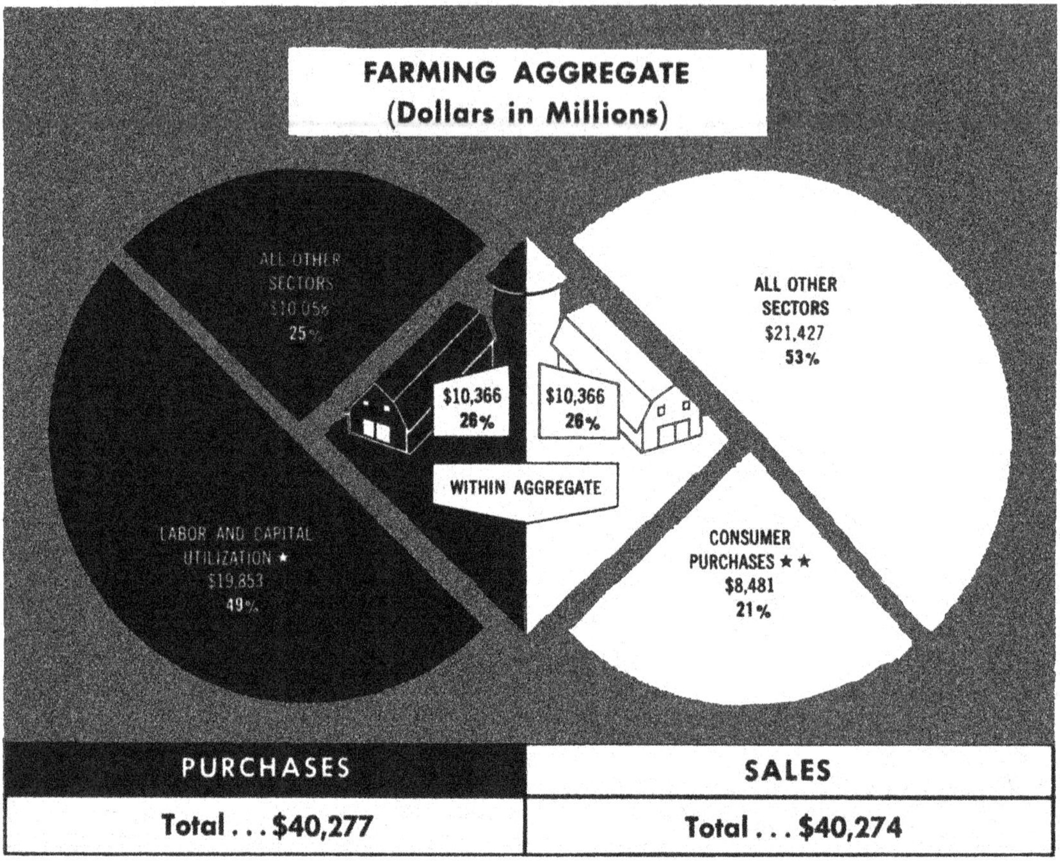

FARMING AGGREGATE
(Dollars in Millions)

ALL OTHER SECTORS
$10,058
25%

LABOR AND CAPITAL UTILIZATION ★
$19,853
49%

$10,366
26%

$10,366
26%

WITHIN AGGREGATE

ALL OTHER SECTORS
$21,427
53%

CONSUMER PURCHASES ★ ★
$8,481
21%

PURCHASES	SALES
Total . . . $40,277	Total . . . $40,274

★ Plus Government Services.

★ ★ Plus Gross Private Capital Formation, Government Purchases, and Inventory Changes.

the factor payment unit, consisting of Labor and Capital Utilization and Government Services. These factor payments accounted for 49% of all farm purchases in 1947, totaling $20 billion, of which $15.5 billion were for such cost items as depreciation, interest, and managerial services.[27] The relatively sizable capital payments by the Farming Aggregate are consistent with the large fixed capital structure portrayed in the agribusiness balance sheets in Chapter 2. The second general category of purchases by farmers, in terms of dollar volume, was that involving intratransactions within farming—i.e., purchases by one farmer from another of such items as feed, feeder cattle, and seed. In 1947 this amounted to over $10 billion, or 26% of all farming purchases. As indicated in Chapter 1, such transactions were relatively much higher prior to the advent of mechanization and technology, when animals were the principal source of power. The remaining 25% of farm purchases were from the All Other Sectors, details of which were presented in the previous section of this chapter relating the Agribusiness Flow Chart to the Input-Output Chart (Exhibit 2).

With respect to sales by the Farming Aggregate, the largest outlets for farm products were the food and fiber processing industries of the economy.

[27] Capital and labor costs are separated out for the Farming Aggregate in Table A-36, Appendix II. A clear picture of the labor-capital relationship is made difficult by the fact that in U.S.D.A. data, part of what would be allocated as a labor and management cost in terms of ordinary business accounting is combined with return to capital as operators' net income in the Farming Aggregate. This is due to the fact that farm owner-operators provide much of the labor force as well as entrepreneurial activity for the Farming Aggregate.

Exhibit 5. Internal Purchases by and Sales of the Farming Aggregate: 1947

GROSS PURCHASES AND SALES

(Millions of dollars)

(a)	SECTOR	1	2	3	4	5	6	7	8	9	10	SALES (b) TOTAL (c)	SALES (b) % (d)	PURCHASES (e) TOTAL (c)	PURCHASES (e) % (d)
1	MEAT ANIMALS AND PRODUCTS (1)	938 938		 130	 3841	 8		 3	 80	 24	*	938	10	5024	51
2	POULTRY AND EGGS (2)		308 308		 1279				 *	 3		308	8	1590	41
3	FARM DAIRY PRODUCTS (3)	130			 1755	 8				 31		130	3	1794	35
4	FOOD GRAINS AND FEED CROPS (4)	3841	1279	1755	817 817	33	20	33	85	63 438		7926	72	1255	11
5	COTTON (5)	8		8	 33	27 27				 108		43	2	168	8
6	TOBACCO (6)				20					 *				20	2
7	OIL-BEARING CROPS (7)	3			 33			93 93		 24		96	9	150	14
8	VEGETABLES AND FRUITS (8)	80	*		 85				121 121	 32		201	5	238	6
9	ALL OTHER AGRICULTURE (9)	24	3	31	438 63	108	*	24	32	61 61		721	37	124	6
10	FISHING, HUNTING, AND TRAPPING (10)	 *									3 3	3	1	3	1
11	TOTAL											10366	26	10366	26

* Less than $500,000; note that these amounts are not included in the total.

(a) Sector code numbers refer to the code numbers of Exhibit 2.

(b) Each sector's sales figures refer to inputs to the Farming Aggregate.

(c) "Total" figures refer to combined sales and purchases of the respective sector of the Farming Aggregate.

(d) Columns refer to the sales and purchases of each sector in the Farming Aggregate as a percentage of the total sales and purchases of the respective sectors in the national economy.

(e) Each sector's purchases refer to outputs of the Farming Aggregate.

Notes: Column and row numbers at the top and left margin refer to the same sectors. Within each square, reading from left to right, the upper and lower figures indicate sales and purchases respectively.

Sectors listed on the left represent the originating source; i.e., these sectors sell to and purchase from those similarly numbered across the top.

Sales to these industries accounted for $21 billion, or 53% of total farm sales in 1947. (See All Other Sectors in Exhibit 4). This broad category is important because most farm products require processing before final consumption. In fact, only about $8 billion of goods and services moved directly to consumers without intermediate processing.[28] As indicated above, farmers also sold 26% of their output to other farmers.

[28] This $8 billion total also includes the other end-factor purchases, such as Government Purchases, Gross Private Capital Formation, minus Inventory Changes.

The relationships of both sales and purchases, sector by sector, are illustrated in Tables 6 and 7.[29] The dominance of capital inputs already has been noted. However, the significance of purchases and

[29] Tables 6 and 7 are derived directly from the Agribusiness input-output matrix (Exhibit 2). In Exhibit 2, if the reader traces down column 11, the Farming Aggregate, the reader is able to ascertain the distribution of the sources of Farming Aggregate purchases, and this distribution is arranged in dollar importance and percentage of total purchases in Table 6. Similarly, in Exhibit 2, row 11, the reader can ascertain the distribution of Farming Aggregate sales, and this distribution is arranged in dollar importance and percentage of total sales in Table 7.

TABLE 6. FARMING AGGREGATE PURCHASES: 1947

Sellers	Rank within Aggregate	Gross Purchases Dollars (Millions)	Gross Purchases % of Total
Labor and Capital Utilization	1	$19,011	47.2%
Food Grains and Feed Crops	2	7,926	19.7
Real Estate and Rentals	3	2,393	5.9
Grain Mill Products	4	2,263	5.6
Meat Animals and Products	5	938	2.3
Government Services (Taxes, etc.)	6	842	2.1
Retail Trade	7	829	2.1
All Other Agriculture	8	721	1.8
Wholesale Trade	9	562	1.4
Petroleum Products	10	450	1.1
Trucking	11	447	1.1
Railroads	12	435	1.1
Fertilizers	13	415	1.0
Automobile and Other Repair Services	14	312	0.8
Poultry and Eggs	15	308	0.8
Maintenance and Construction	16	234	0.6
Other Chemical Industries	17	221	0.6
Vegetable Oils	18	206	0.5
Vegetables and Fruits	19	201	0.5
Wood Containers and Cooperage	20	136	0.3
Farm Dairy Products	21	130	0.3
Tires, Tubes, and Other Rubber Products	22	122	0.3
Oil-Bearing Crops	23	96	0.2
Water and Other Transportation	24	79	0.2
Electric Light and Power	25	56	0.1
Jute, Linens, Cord, and Twine	26	53	0.1
House Furnishings and Other Nonapparel	27	45	0.1
Cotton	28	43	0.1
Miscellaneous Food Products	29	37	0.1
Warehouse and Storage	30	27	0.1
Sugar	31	17	*
Tin Cans and Other Tins	32	16	*
Other Fuels (Coal, Coke, and Gas)	33	15	*
Glass	34	14	*
Alcoholic Beverages	34	14	*
Spinning, Weaving, and Dyeing	36	11	*
Advertising	37	8	*
Fishing, Hunting, and Trapping	38	3	*
Meat Packing and Wholesale Poultry	38	3	*
Animal Oils	38	3	*
Paper and Board Mills and Converted Paper Products	41	2	*
Imports (Noncompetitive)	41	2	*
Metal Container Materials and Cork Products	43	1	*
All Other Industries	—	630	1.6
Total	—	$40,277	100.0%

* Less than 0.05%.
SOURCE: Derived from data published by the Bureau of Labor Statistics, Division of Interindustry Economics.

sales within the Farming Aggregate needs further emphasis. As one might expect, most of these transactions relate to the sectors labeled Meat Animals and Products and All Other Agriculture.

This significance is best traced in Exhibit 5, which is obtained directly from Exhibit 2. Exhibit 5 is comprised of the first ten columns and rows of the agribusiness matrix (Exhibit 2) and

TABLE 7. FARMING AGGREGATE SALES: 1947

Buyers	Rank within Aggregate	Gross Sales: Dollars (Millions)	Gross Sales: % of Total
Consumer Purchases (Households)	1	$9,815	24.4%
Meat Packing and Wholesale Poultry	2	8,406	20.9
Meat Animals and Products	3	5,024	12.5
Grain Mill Products	4	2,496	6.2
Spinning, Weaving, and Dyeing	5	1,984	4.9
Processed Dairy Products	6	1,977	4.9
Farm Dairy Products	7	1,794	4.4
Poultry and Eggs	8	1,590	3.9
Food Grains and Feed Crops	9	1,255	3.1
Vegetable Oils	10	1,106	2.7
Canning, Preserving, and Freezing	11	871	2.2
Eating and Drinking Places	12	865	2.1
Tobacco Manufactures	13	783	1.9
Exports (minus Competitive Imports)	14	696	1.7
Miscellaneous Food Products	15	672	1.7
Alcoholic Beverages	16	348	0.9
Apparel	17	257	0.6
Vegetables and Fruits	18	238	0.6
Saw Mills, Planing, and Veneer Mills	19	188	0.5
Government Purchases	20	181	0.5
Sugar	21	180	0.4
Cotton	22	168	0.4
Oil-Bearing Crops	23	150	0.4
All Other Agriculture	24	124	0.3
Medical, Dental, and Other Professional Services	25	113	0.3
Special Textile Products	26	70	0.2
Bakery Products	27	58	0.1
Leather and Leather Goods	28	51	0.1
Gum and Wood Chemicals	29	39	0.1
Gross Private Capital Formation	30	21	0.1
Tobacco	31	20	0.1
Jute, Linens, Cord, and Twine	31	20	0.1
Animal Oils	33	16	*
Paints and Allied Products	34	15	*
Soap and Related Products	35	6	*
Fishing, Hunting, and Trapping	36	3	*
All Other Industries	—	210	0.5
Inventory Changes	—	—1,536	—3.8
Total	—	$40,274	100.0%

* Less than 0.05%.

SOURCE: Derived from data published by the Bureau of Labor Statistics, Division of Interindustry Economics.

therefore includes each sector's purchases and sales within the Farming Aggregate. In Exhibit 5 these purchases and sales are combined in one block and the chart is read only from left to right. For example, following across row 4, Food Grains and Feed Crops, the reader notes that the amount in the upper portion of each block refers to the sales of the sector at the left to the sectors numbered at the top of the chart; and the amount in the lower portion of each block refers to the purchases by the sector named at the left from the corresponding numbered sectors named at the top. With this in mind, following across row 4 the reader notes the following relations: Food Grain and Feed Crops sold $3,841 million to sector 1, the Meat Animals and Products sector; $1,279 million to sector 2, Poultry and Eggs; $1,755 million to sector 3, Farm Dairy Products; $817 million of goods and services within its own sector (4) [this means that it also purchased $817 million of goods and services from itself]; $33 million to sector 5, Cotton; $20 million to sector 6, Tobacco; $33 million to sector 7, Oil-Bearing Crops; $85 million to sector 8, Vegetables and Fruits; $63 million to sector 9, All Other Agriculture, and in turn purchased $438 million from sector 9. The Food Grains and Feed Crops sector did not sell to sector 10 (Fishing, Hunting, and Trapping) nor did it purchase from this sector.

The total sales of the Food Grains and Feed Crops sector to the Farming Aggregate (row 4 in the Sales column) in 1947 was $7,926 million; and this represented 72% of all the sales of the Food Grains and Feed Crops sector to the total U.S. economy in 1947. The total purchases of the Food Grains and Feed Crops sector from the Farming Aggregate (row 4 in the Purchases column) was $1,255 million which represents 11% of all the purchases made by the Food Grains and Feed Crops sector from the total economy in 1947.

Similar analysis could be made from Exhibit 5 for the remaining nine sectors that form the Farming Aggregate. Together the ten sectors of the Farming Aggregate sold $10,366 million of goods and services to each other (row 11), which comprised 26% of the total sales of the Farming Aggregate in the U.S. economy. These ten sectors also purchased $10,366 million from each other (row 11), in 1947, which comprised 26% of the total purchases of the Farming Aggregate from the U.S. economy.

Exhibit 5 indicates the extent to which farmers growing feed crops in excess of their own need sell the excess to other farmers. In a similar fashion, farmers producing feeder stock (cattle, hogs, or sheep) commonly sell them to other farmers for use as foundation animals or for fattening. The significance of purchases of this type is indicated by the fact that they account for 26% of all purchases by farmers. Because of intrasector transactions, progress in farming is dependent upon the continued improvement of technology within agriculture as well as within the business firms which manufacture farm supplies. This intrasector activity also denotes the degree of specialization that has taken place within agriculture, some farmers being principally livestock growers and others producers of feed. Hence, intra-aggregate activity indicates both interrelationships and specialization.

Table 7 indicates important specific markets for farm products. As already stated, the industries comprising the Food Processing Aggregate, dominated by the Meat Packing and Wholesale Poultry sector, are the largest outlets for farm products, accounting for about 42% of total farm output[30] in 1947. Another important outlet is the Fiber Processing Aggregate, although farm sales to textile and fiber processing industries were smaller than those to food industries.

The Food Processing Aggregate

The Food Processing Aggregate both purchased and sold about $42 billion worth of goods and services in 1947 (see Exhibit 6). The pie chart indicates that Food Processing procured 56% of its inputs by purchases from other intermediate sectors; 24% by means of Labor and Capital Utilization (wages, salaries, depreciation), of which approximately $4 billion was wages and salaries and $5 billion depreciation and similar costs;[31] and the remaining 20% through intra-aggregate transactions.

The most important market for the goods and services produced by the Food Processing Aggregate is the American consumer. As indicated in Exhibit 6,[32] this market amounted to over $24 billion, or 59% of all sales by Food Processing. Also worthy of note is the fact that the industries comprising the Food Processing Aggregate satisfied about 20% of their requirements by purchases from one another.

[30] Table 7 shows the individual rank and percentage of outlets as taken from Exhibit 2; adding the percentages together, one derives that 42% of total farm output is sold to the Food Processing Aggregate.

[31] See Table A-37, Appendix II, which is included to indicate the breakdown between labor costs and other costs in the Labor and Capital Utilization row in the Agribusiness Matrix. To do this the authors have used the basic data in the Interindustry Study. However, this breakdown is not complete enough to include it in the over-all construction of the matrix and is used here and in other parts of the study only to give the reader a more complete descriptive picture of the important segments that form the agribusiness entity.

[32] Included in Consumer Purchases were the following sectors: Gross Private Capital Formation; Government Purchases; and Inventory Changes.

EXHIBIT 6. PURCHASES BY AND SALES OF THE FOOD PROCESSING AGGREGATE: 1947

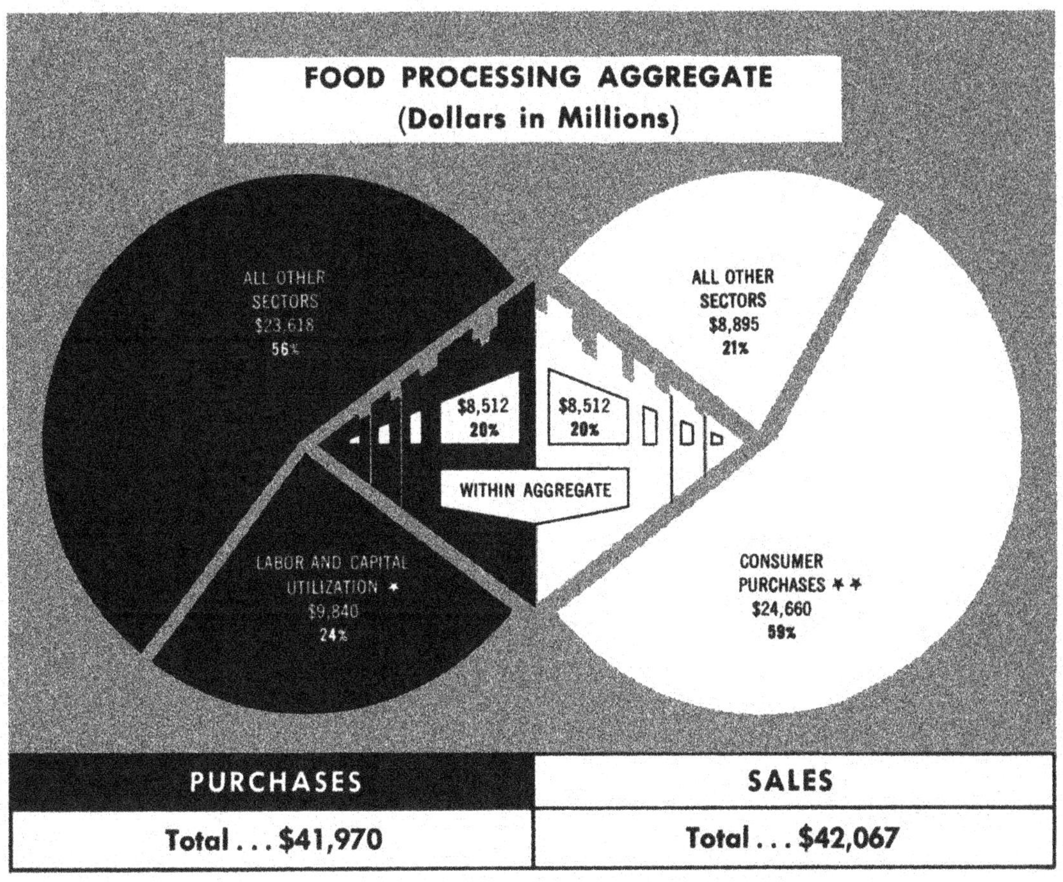

★ Plus Government Services.

★ ★ Plus Gross Private Capital Formation, Government Purchases, and Inventory Changes.

Tables 8 and 9, derived from Exhibit 2, highlight specific purchase and sales relationships of the Food Processing Aggregate to all the sectors and aggregates in the Input-Output Chart. As one might expect, the major sectors supplying the Food Processing Aggregate, with the exception of Labor and Capital Utilization, are located in the Farming Aggregate. These farming sectors sold almost $17 billion, or 42%, of their output of goods and services to Food Processing in 1947. In turn, Food Processing sold $2.5 billion of goods and services to the Farming Aggregate. Purchases from and sales to the Farming Aggregate by food processors were as shown at the top of page 41.

From this tabulation one is able to discern the reciprocity that exists between the Food Processing and Farming Aggregates. In 1947 Food Processing purchased most of the output of the Meat Animals and Products, Oil-Bearing Crops, and Tobacco[34] sectors, reflecting the fact that the commodities of these sectors require a great deal of processing before reaching the final consumer. In contrast, the products of Poultry and Eggs, Vegetables and Fruits, and Farm Dairy Products are bought by the final consumer without extensive change through processing. Although products of Food Grains and Feed Crops require a considerable amount of processing, most of them are used on farms to feed meat animals. Food processing

[34] Tobacco is included here among purchases by the Food Processing Aggregate to be consistent with the arrangement in Exhibit 2.

TABLE 8. FOOD PROCESSING AGGREGATE PURCHASES: 1947

Sellers	*Rank within Aggregate*	*Gross Purchases* Dollars (Millions)	% of Total
Labor and Capital Utilization	1	$8,576	20.4%
Meat Animals and Products	2	8,138	19.4
Food Grains and Feed Crops	3	2,974	7.1
Farm Dairy Products	4	1,965	4.7
Miscellaneous Food Products	5	1,654	4.0
Sugar	6	1,270	3.0
Grain Mill Products	7	1,268	3.0
Government Services (Taxes, etc.)	8	1,264	3.0
Vegetable Oils	9	1,196	2.9
Oil-Bearing Crops	10	934	2.2
Tobacco Manufactures	11	874	2.1
Vegetables and Fruits	12	853	2.0
Imports (Noncompetitive)	13	821	2.0
Meat Packing and Wholesale Poultry	14	782	1.9
Tobacco	15	777	1.9
Paper and Board Mills and Converted Paper Products	16	692	1.7
Railroads	17	614	1.5
Advertising	18	599	1.4
Wholesale Trade	19	560	1.3
Poultry and Eggs	20	540	1.3
Processed Dairy Products	21	522	1.2
Tin Cans and Other Tins	22	506	1.2
Alcoholic Beverages	23	479	1.1
Other Chemical Industries	24	380	0.9
Trucking	25	339	0.8
All Other Agriculture	26	304	0.7
Cotton	27	291	0.7
Glass	28	256	0.6
Animal Oils	29	236	0.6
House Furnishings and Other Nonapparel	30	209	0.5
Canning, Preserving, and Freezing	31	193	0.5
Automobile and Other Repair Services	32	156	0.4
Fishing, Hunting, and Trapping	33	137	0.3
Electric Light and Power	34	134	0.3
Metal Container Materials and Cork Products	35	127	0.3
Wood Containers and Cooperage	36	109	0.3
Other Fuels (Coal, Coke, and Gas)	37	102	0.2
Real Estate and Rentals	38	96	0.2
Maintenance and Construction	39	88	0.2
Petroleum Products	40	69	0.1
Tires, Tubes, and Other Rubber Products	41	57	0.1
Water and Other Transportation	41	57	0.1
Warehouse and Storage	43	51	0.1
Retail Trade	44	43	0.1
Bakery Products	45	38	0.1
Leather and Leather Goods	46	7	*
Apparel	47	6	*
Spinning, Weaving, and Dyeing	48	4	*
Fertilizers	48	4	*
All Other Industries	—	619	1.5
Total	—	$41,970	100.0%

* Less than 0.05%.

SOURCE: Derived from data published by the Bureau of Labor Statistics, Division of Interindustry Economics.

TABLE 9. FOOD PROCESSING AGGREGATE SALES: 1947

Buyers	*Rank within Aggregate*	*Gross Sales* Dollars (Millions)	*% of Total*
Consumer Purchases (Households)	1	$23,728	56.4%
Eating and Drinking Places	2	3,535	8.4
Miscellaneous Food Products	3	1,990	4.7
Bakery Products	4	1,458	3.5
Poultry and Eggs	5	1,309	3.1
Grain Mill Products	6	1,025	2.4
Tobacco Manufactures	7	888	2.1
Sugar	8	685	1.6
Farm Dairy Products	9	650	1.5
Government Purchases	10	604	1.4
Processed Dairy Products	11	530	1.3
Soap and Related Products	12	529	1.3
Alcoholic Beverages	13	517	1.2
Exports (minus Competitive Imports)	14	500	1.2
Leather and Leather Goods	15	453	1.1
Meat Animals and Products	16	446	1.1
Canning, Preserving, and Freezing	17	395	0.9
Animal Oils	18	386	0.9
Meat Packing and Wholesale Poultry	19	345	0.8
Inventory Changes	20	328	0.8
Vegetable Oils	21	293	0.7
Medical, Dental, and Other Professional Services	22	244	0.6
Paints and Allied Products	23	199	0.5
Food Grains and Feed Crops	24	77	0.2
Spinning, Weaving, and Dyeing	25	53	0.1
House Furnishings and Other Nonapparel	26	50	0.1
Special Textile Products	27	28	0.1
Vegetables and Fruits	28	22	0.1
All Other Agriculture	29	15	*
Fishing, Hunting, and Trapping	30	11	*
Tires, Tubes, and Other Rubber Products	31	9	*
Cotton	32	6	*
Oil-Bearing Crops	32	6	*
Apparel	34	5	*
Jute, Linens, Cord, and Twine	35	4	*
Tobacco	36	1	*
Gum and Wood Chemicals		†	**
All Other Industries	—	743	1.8
Total	—	$42,067	100.0%

* Less than 0.05%.
** Less than 0.05%; note that this amount is not included in the total.
† Purchases of less than $500,000; note that this amount is not included in the total.
SOURCE: Derived from data published by the Bureau of Labor Statistics, Division of Interindustry Economics.

sales to the Farming Aggregate consist primarily of sales to the Poultry and Eggs sector since the latter consumes a major share of the mixed feed sold to the farmer.

Because the Food Processing Aggregate is dependent on the Farming Aggregate for its basic materials, its development is affected by the supply, quality, and uniformity of farm products as they flow to market. Also, because numerous by-products of Food Processing are sold back to the Farming Aggregate in the form of food and feed, the supply, quality, and uniformity of these items have an important bearing on the Farming Aggregate.

Exhibit 7, in which purchases and sales are arranged in one block for each sector of the aggre-

EXHIBIT 7. INTERNAL PURCHASES BY AND SALES OF THE FOOD PROCESSING AGGREGATE: 1947

GROSS PURCHASES AND SALES

(Millions of dollars)

(a)	SECTOR	12	13	14	15	16	17	18	19	20	21	22	SALES (b) TOTAL (c)	SALES (b) % (d)	PURCHASES (e) TOTAL (c)	PURCHASES (e) % (d)
12	MEAT PACKING AND WHOLESALE POULTRY (21)	176 176	12 11	38 24	32 3	109 1	111 55	* 5	3 *		1 20	300 50	782	7	345	3
13	PROCESSED DAIRY PRODUCTS (22)	11 12	356 356	2 6	15 1	94 5	44 83	* 59	* 8		*	*	522	14	530	15
14	CANNING, PRESERVING, AND FREEZING (23)	24 38	6 2	43 43	14 21	48 4	41 152	* 79	6 1		* 55	11	193	7	395	14
15	GRAIN MILL PRODUCTS (24)	3 32	1 15	21 14	367 367	699 5	92 98	* 31	54 51		30 286	1 126	1268	24	1 025	19
16	BAKERY PRODUCTS (25)	1 109	5 94	4 48	5 699	10 10	12 370	* 109	1 *		* 19	*	38	1	1 458	44
17	MISC. FOOD PRODUCTS (26)	55 111	83 44	152 41	98 92	370 12	656 656	* 291	33 6	10	178 719	19 18	1 654	25	1 990	30
18	SUGAR (27)	5 *	59 *	79 *	31 *	109 *	291 *	685 685	7 *	4	*	*	1270	108	685	58
19	ALCOHOLIC BEVERAGES (28)	* 3	8 *	1 6	51 54	 1	6 33	* 7	413 413		*	*	479	18	517	19
20	TOBACCO MANUFACTURES (29)						 10	 4		874 874			874	34	888	35
21	VEGETABLE OILS (59)	20 1	 *	55 *	286 30	19 *	719 178	 *	 *		83 83	14 1	1196	62	293	15
22	ANIMAL OILS (60)	50 300	 *	 11	126 1	 *	18 19	 *	 *		1 14	41 41	236	30	386	50
23	TOTAL												8512	20	8512	20

* Less than $500,000; note that these amounts are not included in the total.

(a) Sector code numbers refer to the code numbers of Exhibit 2.

(b) Each sector's sales figures refer to inputs to the Food Processing Aggregate.

(c) "Total" figures refer to combined sales and purchases of the respective sector of the Food Processing Aggregate.

(d) Columns refer to the sales and purchases of each sector in the Food Processing Aggregate as a percentage of the total sales and purchases of the respective sectors in the national economy. Sugar sold 108% of its domestic production to the Food Processing Aggregate. Because imports are included, the percentage of sales totals more than 100% of domestic production.

(e) Each sector's purchases refer to outputs of the Food Processing Aggregate.

NOTES: Column and row numbers at the top and left margin refer to the same sectors. Within each square, reading from left to right, the upper and lower figures indicate sales and purchases respectively.

Sectors listed on the left represent the originating source; i.e., these sectors sell to and purchase from those similarily numbered across the top.

gate, as in Exhibit 5, indicates the purchases and sales among sectors of the Food Processing Aggregate. From this, one may ascertain the flow patterns of the 20% of the total sales and purchases which occur within this aggregate.[35] The magnitude of this interaction points up the opportunity for internal improvement through industry-wide cooperation and policy integration.

[35] Exhibit 7 describes the sector flow within the Food Processing Aggregate just as Exhibit 5 did for the Farming Aggregate. Its relation to the matrix is handled in the same manner as Exhibit 5. (See pages 34 to 36.)

The Fiber Processing Aggregate

This aggregate, whose transactions totaled $26 billion in 1947, is analyzed by means of the same techniques employed in the previous section. The pie chart (Exhibit 8) reveals that the largest inputs were in the form of labor and capital utilization. From Table A–38, Appendix II, one can estimate that approximately 64% of these payments were in the form of wages and salaries and the remainder in capital charges (depreciation, etc.). The second major source of inputs stemmed from intra-aggregate transactions (32% of all

Purchases from and Sales to the Farming Aggregate by the Food Processing Aggregate, 1947[33]

PURCHASES FROM FARMING: *Farming Sectors*	*In Millions*	*% of Total Outputs of Each Farming Sector Purchased by Food Processing*
Meat Animals and Products	$8,138	83%
Food Grains and Feed Crops	2,974	27
Farm Dairy Products	1,965	39
Oil-Bearing Crops	934	88
Vegetables and Fruits	853	21
Tobacco	777	88
Poultry and Eggs	540	14
All Other Agriculture	304	16
Cotton	291	13
Fishing, Hunting, and Trapping	137	34
Total	$16,913	42%

SALES TO FARMING: *Farming Sectors*	*In Millions*	*% of Total Inputs of Each Farming Sector Procured from Food Processing*
Poultry and Eggs	$1,309	34%
Farm Dairy Products	650	13
Meat Animals and Products	446	5
Food Grains and Feed Crops	77	1
Vegetables and Fruits	22	1
All Other Agriculture	15	1
Fishing, Hunting, and Trapping	11	3
Cotton	6	*
Oil-Bearing Crops	6	*
Tobacco	1	*
Total	$2,543	6%

[33] Derived from Tables 8 and 9 and Exhibit 2.

* Less than 0.05%.

EXHIBIT 8. PURCHASES BY AND SALES OF THE FIBER PROCESSING AGGREGATE: 1947

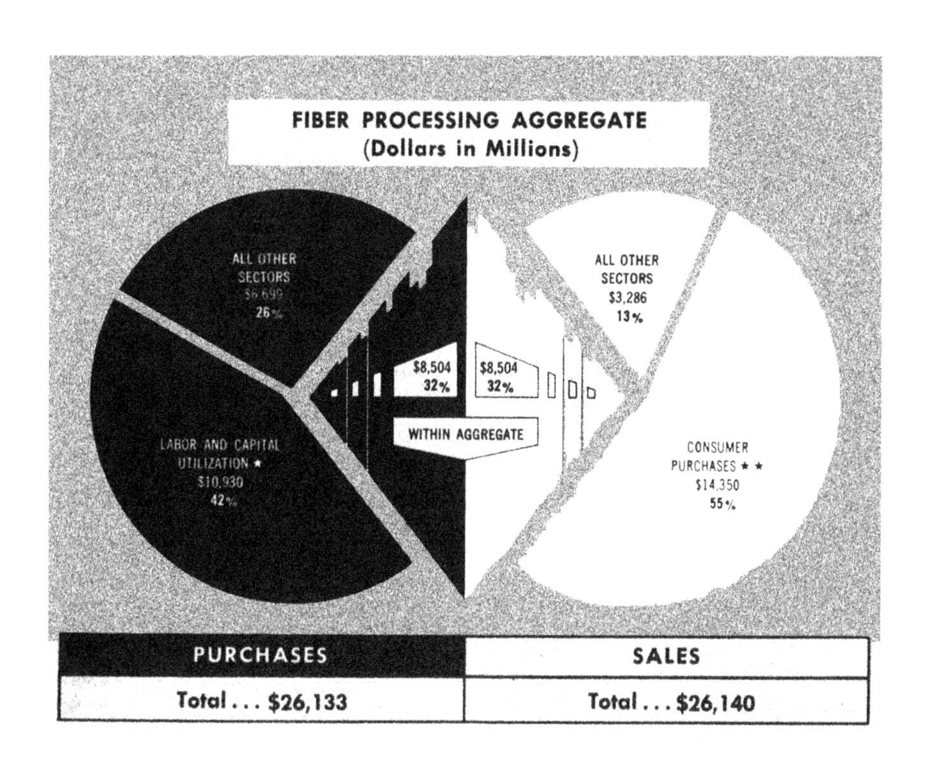

★ Plus Government Services.

★ ★ Plus Gross Private Capital Formation, Government Purchases, and Inventory Changes.

TABLE 10. FIBER PROCESSING AGGREGATE PURCHASES: 1947

Sellers	*Rank within Aggregate*	*Gross Purchases* Dollars (Millions)	% of Total
Labor and Capital Utilization	1	$9,790	37.5%
Spinning, Weaving, and Dyeing	2	5,138	19.7
Apparel	3	1,770	6.8
Cotton	4	1,505	5.8
Leather and Leather Goods	5	1,203	4.6
Government Services (Taxes, etc.)	6	1,140	4.4
Other Chemical Industries	7	1,008	3.9
Wholesale Trade	8	654	2.5
Meat Animals and Products	9	628	2.4
Meat Packing and Wholesale Poultry	10	484	1.9
Fishing, Hunting, and Trapping	11	241	0.9
Paper and Board Mills and Converted Paper Products	12	211	0.8
Imports (Noncompetitive)	13	195	0.7
Railroads	14	193	0.7
Special Textile Products	15	181	0.7
Advertising	16	160	0.6
Real Estate and Rentals	17	141	0.5
House Furnishings and Other Nonapparel	18	131	0.5
Electric Light and Power	19	113	0.4
Tires, Tubes, and Other Rubber Products	20	110	0.4
Trucking	21	87	0.3
Jute, Linens, Cord, and Twine	22	65	0.2
Vegetable Oils	23	61	0.2
Maintenance and Construction	24	53	0.2
Other Fuels (Coal, Coke, and Gas)	25	46	0.2
Petroleum Products	26	35	0.1
Miscellaneous Food Products	26	35	0.1
Warehouse and Storage	28	27	0.1
Water and Other Transportation	29	25	0.1
Automobile and Other Repair Services	30	20	0.1
Wood Containers and Cooperage	30	20	0.1
Canvas Products	32	16	0.1
Animal Oils	33	13	*
Metal Container Materials and Cork Products	34	12	*
All Other Agriculture	35	8	*
Glass	36	2	*
Retail Trade	37	1	*
All Other Industries	—	611	2.3
Total	—	$26,133	100.0%

* Less than 0.05%.

SOURCE: Derived from data published by the Bureau of Labor Statistics, Division of Interindustry Economics.

purchases and sales), and the third from other processing industries.

On the output or sales side, Exhibit 8 indicates that 55% of all sales move directly to final private consumers or to government purchases, inventory changes, and gross private capital formation in the amount of $14,350 million. Thirty-two per cent of all output is sold within the Fiber Processing Aggregate, indicating the complexity of operations within this aggregate. Exhibit 9 demonstrates in a more exact manner the interrelationships within the Fiber Processing Aggregate similar to the analysis of the Food Processing Aggregate and Farming Aggregate in Exhibits 7 and 5 respectively.

The relationship of inputs to outputs of this aggregate are shown in more detail in Tables 10 and 11. Purchases by the Fiber Aggregate from

TABLE 11. FIBER PROCESSING AGGREGATE SALES: 1947

Buyers	*Rank within Aggregate*	*Gross Sales* *Dollars (Millions)*	*% of Total*
Consumer Purchases (Households)	1	$13,425	51.3%
Apparel	2	4,967	19.0
Leather and Leather Goods	3	1,236	4.7
Exports (minus Competitive Imports)	4	1,094	4.2
Spinning, Weaving, and Dyeing	5	1,048	4.0
House Furnishings and Other Nonapparel	6	961	3.7
Inventory Changes	7	460	1.8
Tires, Tubes, and Other Rubber Products	8	444	1.7
Government Purchases	9	426	1.6
Special Textile Products	10	201	0.8
Grain Mill Products	11	188	0.7
Canvas Products	12	50	0.2
Jute, Linens, Cord, and Twine	13	41	0.2
Gross Private Capital Formation	14	39	0.1
Vegetables and Fruits	15	34	0.1
All Other Agriculture	16	32	0.1
Medical, Dental, and Other Professional Services	17	27	0.1
Eating and Drinking Places	18	21	0.1
Fishing, Hunting, and Trapping	19	20	0.1
Sugar	20	11	*
Vegetable Oils	20	11	*
Saw Mills, Planing, and Veneer Mills	22	10	*
Food Grains and Feed Crops	23	9	*
Tobacco	24	8	*
Meat Packing and Wholesale Poultry	25	7	*
Cotton	26	4	*
Miscellaneous Food Products	26	4	*
Animal Oils	28	3	*
Meat Animals and Products	29	1	*
Oil-Bearing Crops	29	1	*
Processed Dairy Products	29	1	*
Canning, Preserving, and Freezing	29	1	*
Soap and Related Products		†	**
Paints and Allied Products		†	**
All Other Industries	—	1,355	5.2
Total	—	$26,140	100.0%

* Less than 0.05%.
** Less than 0.05%; note that these amounts are not included in the total.
† Purchases of less than $500,000; note that these amounts are not included in the total.
SOURCE: Derived from data published by the Bureau of Labor Statistics, Division of Interindustry Economics.

sectors of the Farming Aggregate, as derived from Table 10, are indicated in the tabulation at the bottom of page 44.

From this abbreviated table one may note that although cotton is a relatively small input of Fiber Processing, this aggregate consumes 68% of the cotton output of the Farming Aggregate. Similarly, while Fishing, Hunting, and Trapping supplies less than 1% of all the inputs of Fiber Processing, yet this amounts to 60% of all the output of the Fishing, Hunting, and Trapping sector. One may conclude that the role of the Fiber Processing Aggregate as a market for the above listed farming sectors is greater, quantitatively, than the sectors' role to the Fiber Processing Aggregate as suppliers of raw materials.

Sales of the Fiber Processing Aggregate as shown in detail in Table 11, when related to the raw materials purchases by this aggregrate from the Farming Aggregate, highlight the low relative cost factor of raw fibers in relation to the selling price of fiber end products (in producers' values).

EXHIBIT 9. INTERNAL PURCHASES BY AND SALES OF THE FIBER PROCESSING AGGREGATE: 1947

GROSS PURCHASES AND SALES
(Millions of dollars)

(a)	SECTOR	24	25	26	27	28	29	30	SALES (b) TOTAL (c)	SALES (b) % (d)	PURCHASES (e) TOTAL (c)	PURCHASES (e) % (d)
24	SPINNING, WEAVING, AND DYEING ( )	990 990	104 17	28 22	36	3043	876 17	61 2	5138	63	1048	13
25	SPECIAL TEXTILE PRODUCTS (31)	17 104	77 77	3 18		70	5 2	9	181	22	201	24
26	JUTE, LINENS, CORD, AND TWINE (32)	22 28	18 3	9 9		5	9 1	2	65	25	41	16
27	CANVAS PRODUCTS (33)	 36			7 7	1 *	8 7		16	16	50	52
28	APPAREL (34)	 3043	 70	 5	* 1	1765 1765	4 26	1 57	1770	16	4967	44
29	HOUSE FURNISHINGS (35) AND OTHER NONAPPAREL	17 876	2 5	1 9	7 8	26 4	59 59	19	131	7	961	53
30	LEATHER AND LEATHER GOODS (67-69)	2 61	 9	 2		57 1	 19	1144 1144	1203	32	1236	33
31	TOTAL								8504	32	8504	32

* Less than $500,000; note that these amounts are not included in the total.
(a) Sector code numbers refer to the code numbers of Exhibit 2.
(b) Each sector's sales figures refer to inputs to the Fiber Processing Aggregate.
(c) "Total" figures refer to combined sales and purchases of the respective sector of the Fiber Processing Aggregate.
(d) Columns refer to the sales and purchases of each sector in the Fiber Processing Aggregate as a percentage of the total sales and purchases of the respective sectors in the national economy.
(e) Each sector's purchases refer to outputs of the Fiber Processing Aggregate.

NOTES: Column and row numbers at the top and left margin refer to the same sectors. Within each square, reading from left to right, the upper and lower figures indicate sales and purchases respectively.

Sectors listed at the left represent the originating source; i.e., these sectors sell to aand purchase from those similarly numbered across the top.

Purchases of the Fiber Processing Aggregate from the Farming Aggregate, 1947

Farming Sectors	*Millions*	*% of Total Purchases by Fiber Processing Aggregate*	*% of Total Output of Each Farming Sector Purchased by Fiber Processing*
Cotton	$1,505	5.8%	68%
Meat Animals and Products	628	2.4	6
Fishing, Hunting, and Trapping	241	0.9	60
All Other Agriculture	8	—	—
Total	$2,382	9.1%	6%

For example, the previous abbreviated table indicated purchases of $2,382 million from the Farming Aggregate sectors and Table 11 indicates Fiber Processing final demand consumption as follows:

	Millions
Consumer Purchases (Households)	$13,425
Inventory Changes (increase)	460
Government Purchases	426
Gross Private Capital Formation	39
	$14,350

In addition, $1,094 million of finished products were exported. Comparing the purchases of natural fiber of $2,382 million to these finished fiber product sales, one arrives at a percentage figure of slightly over 15% of the value of finished sales being accounted for by the cost of the natural fibers. In the case of the Food Processing Aggregate, purchases by this aggregate of raw food products from the Farming Aggregate constituted over 67% of the value of the food sales to final consuming sectors. Viewed from the standpoint of processing, these price relationships reflect the more complex operations involved in converting fibers to end products than in the case of foods. This complexity is further indicated by the fact that intra-aggregate activity amounts to 32% of all purchases and sales by this aggregate (see Exhibits 8 and 9). The greater portion of this fiber intra-aggregate activity takes place within the apparel sector as indicated in Exhibit 9 and Tables 10 and 11.

The Secondary Triaggregate

Thus far in this section the authors have examined the general flow of goods and services through each of three aggregates: Farming, Food Processing, and Fiber Processing. These include the major agribusiness sectors shown in the Input-Output Chart, Exhibit 2. For convenience we henceforth will refer to these three aggregates as the Secondary Triaggregate.[36]

Keeping in mind the omission of the Farm Supplies Aggregate as such from this analysis, let us examine the relationship of this Secondary Triaggregate to the individual sectors that comprise it and to the rest of the economy. In order to facilitate this analysis, Exhibit 10 was prepared to illustrate the reciprocity between the sectors of the three aggregates in terms of sales and purchases—both in dollar values and as a percentage of the total for the triaggregate. The data in the three outlined blocks in Exhibit 10 correspond to those in Exhibits 5, 7, and 9, whereas those in the remaining areas consist of interactions among the aggregates. The interdependency of the sectors of this triaggregate is highlighted by the following facts:

(a) Forty-six per cent of all purchases and sales of this secondary triaggregate are made between entities within it, thus indicating the strength of the forces that hold agribusiness together. The nature of these transactions is indicated in the following:

Total Transactions of Secondary Triaggregate (Farming, Food Processing, Fiber Processing)

Secondary Triaggregate Purchased in 1947 (millions of dollars)		*% of Total Output of Each Aggregate Purchased by Secondary Triaggregate*
$29,661	from Farming	73.6%
11,648	from Food Processing	27.7
8,839	from Fiber Processing	33.8
$50,148	Grand Total of Secondary Triaggregate	46%

Secondary Triaggregate Sold in 1947 (millions of dollars)		*% of Total Input per Aggregate Procured from the Secondary Triaggregate*
$13,018	to Farming	32.5%
25,651	to Food Processing	61.1
11,479	to Fiber Processing	43.9
$50,148	Grand Total of Secondary Triaggregate	46%

Relationships of Farming, Food Processing, and Fiber Processing to Each Other and to Total Secondary Triaggregate

Farming Purchased in 1947 (millions of dollars)		*% of Total Farming Purchases*	*% of Total Secondary Triaggregate Purchases*
$10,366	from itself	26%	10%
2,543	from Food Processing	6	2
109	from Fiber Processing	—	—

Farming Sold in 1947 (millions of dollars)		*% of Total Farming Sales*	*% of Total Secondary Triaggregate Sales*
$10,366	to itself	26%	10%
16,913	to Food Processing	42	16
2,382	to Fiber Processing	6	2

Food Processing Purchased in 1947 (millions of dollars)		*% of Total Food Processing Purchases*	*% of Total Secondary Triaggregate Purchases*
$16,913	from Farming	40%	16%
8,512	from itself	20	8
226	from Fiber Processing	1	—

[36] In the analysis of agribusiness in Chapters 1 and 2, the three agribusiness aggregates of Farm Supplies, Farming, and Processing-Distribution were referred to as the Primary Triaggregate. Because of limitations in the sources of data, we do not have a complete enough breakdown of the Farm Supplies Aggregate to include it as such in Exhibit 3. Therefore, for convenience of analysis we are considering the remaining component parts of agribusiness as a new secondary triaggregate consisting of Farming, Food Processing, and Fiber Processing.

EXHIBIT 10. INTERRELATIONSHIPS AMONG SECTORS OF THE FARMING, FOOD PROCESSING, AND FIBER PROCESSING AGGREGATES: 1947

GROSS PURCHASES AND SALES

(Millions of dollars)

	(a)	SECTOR	1	2	3	4	5	6	7	8	9	10	12	13	14	15	16	17	18	19	20	21	22	24	25	26	27	28	29	30	SALES (b) TOTAL (c)	SALES (b) % (d)	PURCHASES (e) TOTAL (c)	PURCHASES (e) % (d)
FARMING AGGREGATE	1	MEAT ANIMALS AND PRODUCTS (1)	938 938		130	3841	8		3	80	24	X	8099			314		39	4			125	3	514	67					47	9704	99	5471	56
	2	POULTRY AND EGGS (2)		306 308		1279				X	3		297	18	27	1296	17	181 5				18									848	22	2899	75
	3	FARM DAIRY PRODUCTS (3)	130			1755	8				31		10	1929		541	20	6 19	13	14		63									2096	41	2444	48
	4	FOOD GRAINS AND FEED CROPS (4)	3841	1279	1755	817 817	33	20	33	85	63 438				4	2466 77		246		259						3			8		10900	99	1341	12
	5	COTTON (5)	8		8	33	27 27				108					6						291		1470 4		17		18			1930	83	178	8
	6	TOBACCO (6)				20					X					1					777			7		1					777	88	29	3
	7	OIL-BEARING CROPS (7)	3			33			93 93		24					25 6	2	92				815							1		1030	97	157	15
	8	VEGETABLES AND FRUITS (8)	80	X		65				121 121	32			20	721	3 16	13	39 6		57									34		1054	26	294	7
	9	ALL OTHER AGRICULTURE (9)	24	3	31	438 63	108	X	24	32	61 61			10	1	15	6	69	180	32	6	X			3	3 29			3	2	1033	53	171	9
	10	FISHING, HUNTING, AND TRAPPING (10)	X									3 3	3		118	2 1		1 7					16			20		239		2	381	94	34	8
FOOD PROCESSING AGGREGATE	12	MEAT PACKING AND WHOLESALE POULTRY (21)	8099	297	10							3	176 176	12 11	38 24	32 3	109 1	111 55	X 5	3 X		1 20	300 50	19	17			6	5 1	443	1269	11	8758	79
	13	PROCESSED DAIRY PRODUCTS (22)		18	1929					20	10		11 12	356 356	2 6	15 1	94 5	44 83	X 59	X 8		X	X	1							522	14	2506	69
	14	CANNING, PRESERVING, AND FREEZING (23)		27		4				721	1	118	24 38	6 2	43 43	14 21	48 4	41 152	X 79	6 1		X 55	11							1	193	7	1267	46
	15	GRAIN MILL PRODUCTS (24)	314	1296	541	77 2466	6	1	6 25	16 3	15	1 2	3 32	1 15	21 14	367 367	699 5	92 98	X 31	54 51		30 286	1 126						184	4	3531	66	3709	68
	16	BAKERY PRODUCTS (25)		17	20				2	13	6		1 109	5 94	4 46	5 699	10 10	12 370	X 109	1 X		X 19	X								38	1	1516	45
	17	MISCELLANEOUS FOOD PRODUCTS (26)	39	5 181	19 6	246			92	6 39	69	7 1	55 111	83 44	152 41	98 92	370 12	656 656	X 291	33 6	10	178 719	19 18	26 1	X				5 2	4 1	1726	26	2666	40
	18	SUGAR (27)	4		13						180		5 X	59 X	79 X	31 X	109 X	291 X	685 685	7 X	4	X	X						11		1287	109	876	74
	19	ALCOHOLIC BEVERAGES (28)			14	259				57	32		X 3	8 X	1 6	51 54	X 1	6 33	X 7	413 413		X	X								495	18	665	32
	20	TOBACCO MANUFACTURES (29)						777			6							10	4		874 874										874	34	1671	65
	21	VEGETABLE OILS (59)	125	18	63		291		815		X		20 1	X	55 X	286 30	19 X	719 178	X	X		83 83	14 1	2	11	4		8	38 10	1 1	1463	76	1410	74
	22	ANIMAL OILS (60)	3									16	50 300	X	11	126 1	X	18 19	X	X		1 14	41 41	6 2					2 1	5	252	32	405	52
FIBER PROCESSING AGR.	24	SPINNING, WEAVING, AND DYEING (30)	514				4 1470	7					19	1				1 26				2	2 6	990 990	104 17	28 22	36	3043	876 17	61 2	5153	64	3085	38
	25	SPECIAL TEXTILE PRODUCTS (31)	67								3		17					X				11		17 104	77 77	3 18		70	5 2	9	181	22	299	36
	26	JUTE, LINENS, CORD, AND TWINE (32)				3	17	1			29 3	20										4		22 28	18 3	8 9		5	9 1	2	118	46	68	26
	27	CANVAS PRODUCTS (33)																						56			7 7	1 X	8		16	16	50	62
	28	APPAREL (34)					18					239	6									5		3043	70	5	X 1	1765 1765	4 26	1 57	1776	18	8229	46
	29	HOUSE FURNISHINGS AND OTHER NONAPPAREL (35)	1			6			1	34	3		1 5			184		2 5	11			10 38	1 2	17 676	2 5	1 9	7 8	26 4	59 59	19	385	21	1011	56
	30	LEATHER AND LEATHER GOODS (67-69)	47								2	2	443		1	4		1 4				1 1	5	2 61	9	2		57 1	19	1144 1144	1210	32	1740	47
		TOTAL OF ABOVE AGGREGATES																													50 148	46	50,148	46

Food Processing Sold in 1947 (millions of dollars)	*% of Total Food Processing Sales*	*% of Total Secondary Triaggregate Sales*
$2,543 to Farming	6%	2%
8,512 to Itself	20	8
593 to Fiber Processing	1	1

Fiber Processing Purchased in 1947 (millions of dollars)	*% of Total Fiber Processing Purchases*	*% of Total Secondary Triaggregate Purchases*
$2,382 from Farming	9%	2%
593 from Food Processing	2	1
8,504 from itself	32	8

Fiber Processing Sold in 1947 (millions of dollars)	*% of Total Fiber Processing Sales*	*% of Total Secondary Triaggregate Sales*
$ 109 to Farming	—	—
226 to Food Processing	1%	—
8,504 to itself	32	8%

(b) Twelve of the 28 sectors of the Secondary Triaggregate had over 70% of their purchases or sales take place within the triaggregate. These sectors are as follows:

PURCHASES BY INDUSTRY SECTORS:

Industry Sectors	*Millions*	*% of Total Input of Each Sector Procured within Secondary Triaggregate*
Poultry and Eggs	$2,899	75%
Meat Packing and Wholesale Poultry	8,758	79
Sugar	876	74
Vegetable Oils	1,410	74

SALES FROM INDUSTRY SECTORS:

Industry Sectors	*Millions*	*% of Total Output of Each Sector Purchased within Secondary Triaggregate*
Meat Animals and Products	$9,704	99%
Food Grains and Feed Crops	10,900	99
Cotton	1,839	83
Tobacco	777	88
Oil-Bearing Crops	1,030	97
Fishing, Hunting, and Trapping	381	94
Sugar	1,287	109*
Vegetable Oils	1,463	76

* Output of the Sugar sector exceeds 100% because of imports added to its domestic production.

Of the agribusiness sectors that purchased 70% or more of their total inputs from within the Secondary Triaggregate, one was a Farming sector[37] (Poultry and Eggs) utilizing a considerable amount of feed crops and mixed feeds, and the other three were Food Processing sectors which in turn alter the form of farm commodities. Other Secondary Triaggregate industry sectors usually require more labor inputs or Farm Supplies Aggregate inputs such as machinery, power, and fertilizer, than commodities, in terms of dollar values.

Of the eight sectors that sold over 70% of their total output within the Secondary Triaggregate, six were Farming sectors whose products require additional processing before final consumption, and two were from the Food Processing Aggregate (Sugar and Vegetable Oils).

In order to round out the picture with respect to agribusiness transactions, let us examine input-

[37] Additional farming sectors as purchases would be represented if Farm Supplies Aggregate data had been available for this analysis.

NOTES TO EXHIBIT 10.

* Less than $500,000; note that these amounts are not included in the total.

(a) Sector code numbers refer to the code numbers of Exhibit 2.

(b) Each sector's sales figures refer to inputs to the composite aggregate of Farming, Food Processing, and Fiber Processing.

(c) "Total" figures refer to combined sales and purchases of the respective sectors for the three aggregates.

(d) Columns refer to the sales and purchases of each sector in the combined Farming, Food Processing, and Fiber Processing Aggregates as a percentage of the total sales and purchases of the respective sectors in the national economy.

(e) Each sector's purchases refer to outputs of the composite aggregate of Farming, Food Processing, and Fiber Processing.

NOTES: Column and row numbers at top and left margin refer to the same sectors. Within each square, reading from left to right, the upper and lower numbers indicate sales and purchases, respectively.

Sectors listed at the left represent the originating source; i.e., these sectors sell to and purchase from those similarly numbered across the top.

output examples drawn from outside the Secondary Triaggregate.

The Tires, Tubes, and Other Rubber Products Sector

Even though this sector is not completely a part of agribusiness, it was chosen for analysis because it is an important supplier to agribusiness industries, generally, and a purchaser of many of their outputs. The same type of derived figures and tables are shown as in our previous examinations, in order to enable the reader to visualize the general characteristics of the Tires, Tubes, and Other Rubber Products sector and its relationship with agribusiness sectors.

The pie chart (Exhibit 11) indicates that the over-all pattern of this sector's inputs and outputs totaled approximately $3 billion in 1947.

Tables 12 and 13 present in more detail the reciprocal relationship between this sector and the Secondary Triaggregate in terms of inputs and outputs. From these tables, the relationship of the Tires, Tubes, and Other Rubber Products sector to the Secondary Triaggregate of Farming, Food Processing, and Fiber Processing is summarized as follows:

Transactions between the Tires, Tubes, and Other Rubber Products Sector and the Secondary Triaggregate,[38] *1947*

PURCHASES FROM THE SECONDARY TRIAGGREGATE SECTORS:

Secondary Triaggregate Sectors	*Millions of Dollars*	*As % of Total Purchases of Tires, Tubes, and Other Rubber Products Sector*	*% of Total Output of Each Secondary Triaggregate Sector Purchased by Tires, Tubes, and Other Rubber Products Sector*
Spinning, Weaving, and Dyeing	$441	14.7%	5.4%
Animals Oils	9	.3	1.2
House Furnishings and Other Nonapparel	3	.1	.2
Totals	$453	15.1%	.4*

* Percentage of total Secondary Triaggregate output purchased by Tires, Tubes, and Other Rubber Products sector.

[38] Derived from Tables 12 and 13 and Exhibit 2.

SALES TO THE SECONDARY TRIAGGREGATE SECTORS:

Secondary Triaggregate Sectors	*Millions of Dollars*	*As % of Total Sales of Tires, Tubes, and Other Rubber Products Sector*	*% of Total Input of Each Secondary Triaggregate Sector Procured from Tires, Tubes, and Other Rubber Products Sector*
Food Grains and Feed Crops	$57	1.9%	.5%
Leather and Leather Goods	55	1.8	1.5
Apparel	34	1.1	.3
House Furnishings and Other Nonapparel	16	.5	.9
Vegetables and Fruits	14	.5	.3
Meat Animals and Products	11	.4	.1
All Other Agriculture	11	.4	.6
Farm Dairy Products	10	.3	.2
Meat Packing and Wholesale Poultry	10	.3	.1
Miscellaneous Food Products	10	.3	.2
Bakery Products	9	.3	.3
Cotton	8	.3	.4
Grain Mill Products	7	.2	.1
Alcoholic Beverages	6	.2	.2
Oil-Bearing Crops	5	.2	.5
Poultry and Eggs	4	.1	.1
Processed Dairy Products	4	.1	.1
Animal Oils	4	.1	.5
Spinning, Weaving, and Dyeing	4	.1	*
Canning, Preserving, and Freezing	3	.1	.1
Tobacco	2	.1	.2
Tobacco Manufactures	2	.1	.1
Vegetable Oils	2	.1	.1
Special Textile Products	1	*	.1
Totals	$289	9.5%	.3%†

* Less than 0.05%.

† Percentage of total Secondary Triaggregate input procured from Tires, Tubes, and Other Rubber Products sector.

This Tires, Tubes, and Other Rubber Products sector is illustrative of numerous industries which do only a part of their total business with agribusiness, but which both procure raw materials from and sell finished products to the Secondary

EXHIBIT 11. PURCHASES BY AND SALES OF THE TIRES, TUBES, AND OTHER RUBBER PRODUCTS SECTOR: 1947

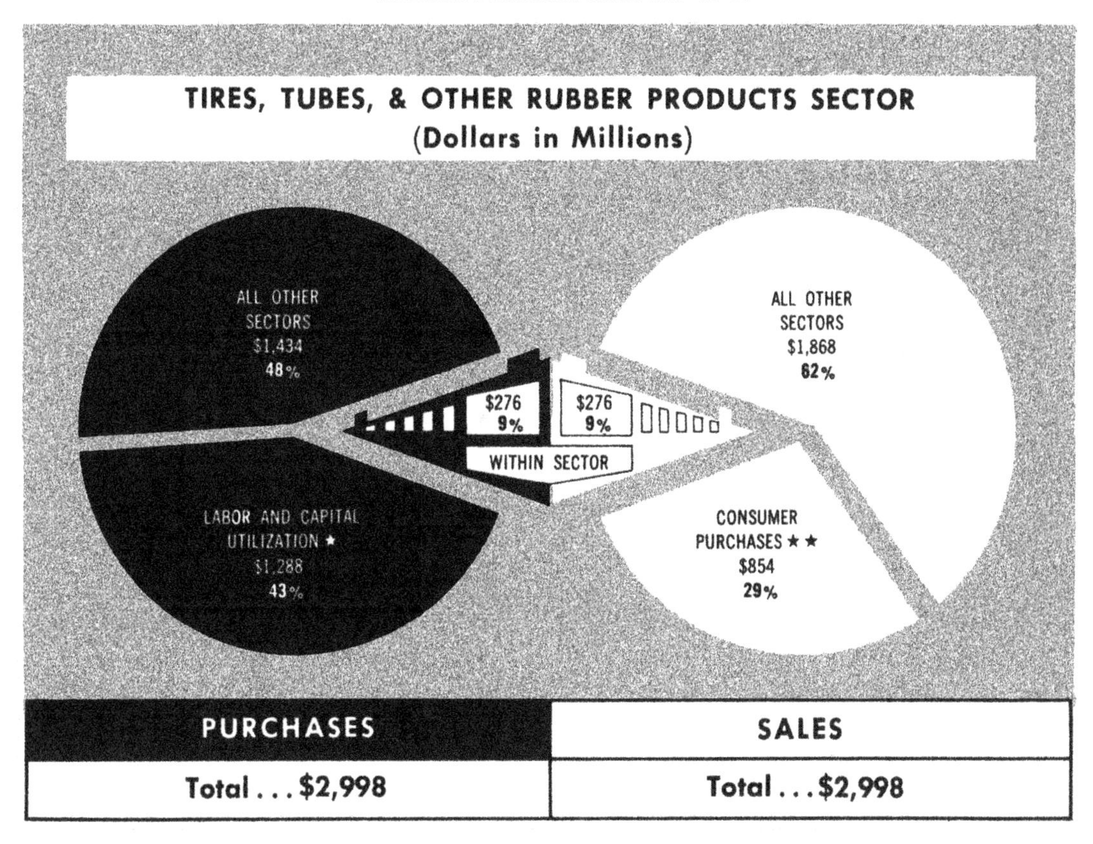

★ Plus Government Services.

★★ Plus Gross Private Capital Formation, Government Purchases, and Inventory Changes.

Triaggregate. The extensive flow of goods and services between such perimeter industries and major agribusiness industries indicates additional magnitudes of interactions.

The All Other Industries Aggregate

This aggregate, as pointed out in the first part of this chapter, is a residual unit composed of those processing sectors which are not specifically designated in the matrix, Exhibit 2. Because all the sectors arranged at the side and top of the matrix are not identical, the All Other Industries Aggregate, side and top of the matrix, differs correspondingly.

As indicated in the tabulation on page 52, the Secondary Triaggregate accounted for only about 1% of either the total purchases or sales of the All Other Industries Aggregate. In turn, the All Other Industries Aggregate accounted for only 2.1% of the sales and 1.7% of the purchases of the Secondary Triaggregate. (However, a few sectors of the Secondary Triaggregate, such as Spinning, Weaving, and Dyeing, do have a significant relation to the All Other Industries Aggregate.)

In summary, the All Other Industries Aggregate is composed of industry sectors that fall into two categories: (1) those industry sectors in the national economy that neither sell to nor buy from agribusiness industry sectors; and (2) those industry sectors that are part of agribusiness but have not been separated out because of lack of adequate data (e.g., the case of farm machinery that has its total output placed in the Gross Private Capital Formation column rather than allocated to

TABLE 12. TIRES, TUBES, AND OTHER RUBBER PRODUCTS SECTOR PURCHASES: 1947

Sellers	Rank within Sector	Gross Purchases Dollars (Millions)	Gross Purchases % of Total
Labor and Capital Utilization	1	$1,168	39.0%
Other Chemical Industries	2	621	20.7
Spinning, Weaving, and Dyeing	3	441	14.7
Tires, Tubes, and Other Rubber Products	4	276	9.2
Government Services (Taxes, etc.)	5	120	4.0
Wholesale Trade	6	58	1.9
Railroads	7	37	1.2
Paper and Board Mills and Converted Paper Products	8	36	1.2
Electric Light and Power	9	25	0.8
Advertising	10	20	0.7
Petroleum Products	11	14	0.5
Other Fuels (Coal, Coke, and Gas)	12	13	0.4
Real Estate and Rentals	13	10	0.3
Animal Oils	14	9	0.3
Wood Containers and Cooperage	14	9	0.3
Maintenance and Construction	16	8	0.3
Automobile and Other Repair Services	17	5	0.2
Trucking	18	4	0.1
House Furnishings and Other Nonapparel	19	3	0.1
Glass	19	3	0.1
Water and Other Transportation	21	2	0.1
Warehouse and Storage	22	1	*
Retail Trade	22	1	*
All Other Industries	—	114	3.8
Total	—	$2,998	100.0%

* Less than 0.05%.

SOURCE: Derived from data published by the Bureau of Labor Statistics, Division of Interindustry Economics.

the respective Farming Aggregate sectors). Because of the All Other Industries Aggregate's heterogeneous nature, we will dispense with the types of description and analysis applied to other aggregates considered in this section.

Final Demand and Factor Payment Sectors

Having analyzed, within the limitations of data available, the relationships of agribusiness sectors to other sectors engaged in processing and the relationship of these to the rest of the economy, let us now turn attention to the factor payment[40] and end-product sectors.[41] The end-product sectors selected are Consumer Purchases (household end-product consumption) and Government Purchases[42] (purchases by public agencies) on the output side of the matrix and their respective counterparts on the input side, factor payments to Labor and Capital Utilization[43] and taxes paid for Government Services.[44] These two sectors have been chosen to indicate the principal users, both private and public, of final agribusiness outputs and the factor payments in the form of labor and capital inputs.

[40] Factor payment sectors refer to those inputs to an industry that come from outside of the processing or intermediate economy—these inputs include noncompetitive imports, taxes, wages, capital costs, managerial costs, profits, etc.

[41] End-product sectors refer to those sectors that purchase finished products directly from each processing sector—these sectors include Exports, Government Purchases, Gross Private Capital Formation, and Consumer Purchases.

[42] Government Purchases is defined in Appendix I.

[43] Depreciation, profits, interest, wages, etc.

[44] Government Services are considered as federal, state, and local government "sales" to the economy for the various functions; i.e., "services" that government performs—defense, social security, fire, health, welfare, etc. See Appendix I.

TABLE 13. TIRES, TUBES, AND OTHER RUBBER PRODUCTS SECTOR SALES: 1947

Buyers	Rank within Sector	Gross Sales: Dollars (Millions)	Gross Sales: % of Total
Consumer Purchases (Households)	1	$731	24.4%
Tires, Tubes, and Other Rubber Products	2	276	9.2
Exports (minus Competitive Imports)	3	168	5.6
Inventory Changes	4	87	2.9
Food Grains and Feed Crops	5	57	1.9
Leather and Leather Goods	6	55	1.8
Apparel	7	34	1.1
Government Purchases	8	29	1.0
House Furnishings and Other Nonapparel	9	16	0.5
Vegetables and Fruits	10	14	0.5
Saw Mills, Planing, and Veneer Mills	11	13	0.4
Meat Animals and Products	12	11	0.4
All Other Agriculture	12	11	0.4
Farm Dairy Products	14	10	0.3
Meat Packing and Wholesale Poultry	14	10	0.3
Miscellaneous Food Products	14	10	0.3
Bakery Products	17	9	0.3
Cotton	18	8	0.3
Medical, Dental, and Other Professional Services	18	8	0.3
Grain Mill Products	20	7	0.2
Gross Private Capital Formation	20	7	0.2
Alcoholic Beverages	22	6	0.2
Soap and Related Products	22	6	0.2
Oil-Bearing Crops	24	5	0.2
Paints and Allied Products	24	5	0.2
Poultry and Eggs	26	4	0.1
Processed Dairy Products	26	4	0.1
Animal Oils	26	4	0.1
Spinning, Weaving, and Dyeing	26	4	0.1
Eating and Drinking Places	26	4	0.1
Canning, Preserving, and Freezing	31	3	0.1
Tobacco	32	2	0.1
Tobacco Manufactures	32	2	0.1
Vegetable Oils	32	2	0.1
Special Textile Products	35	1	*
Gum and Wood Chemicals	35	1	*
Jute, Linens, Cord, and Twine		†	**
Canvas Products		†	**
All Other Industries	—	1,374	45.9
Total	—	$2,998	100.0%

* Less than 0.05%.

**Less than 0.05%; note that these amounts are not included in the total.

† Purchases of less than $500,000; note that these amounts are not included in the total.

SOURCE: Derived from data published by the Bureau of Labor Statistics, Division of Interindustry Economics.

Transactions Between the All Other Industries Aggregate and the Secondary Triaggregate,[39] 1947

PURCHASES FROM THE SECONDARY TRIAGGREGATE SECTORS:

Secondary Triaggregate Sectors	Millions of Dollars	As % of Total Purchases by the All Other Industries Aggregate	% of Total Output of Each Secondary Triaggregate Sector Purchased by the All Other Industries Aggregate
Spinning, Weaving, and Dyeing	$509	.2%	6.3%
House Furnishings and Other Nonapparel	336	.1	18.6
Miscellaneous Food Products	233	.1	3.5
Special Textile Products	175	.1	21.3
Leather and Leather Goods	167	.1	4.5
All Other Agriculture	147	*	7.5
Vegetable Oils	132	*	6.9
Animal Oils	105	*	13.4
Meat Packing and Wholesale Poultry	85	*	.8
Jute, Linens, Cord, and Twine	83	*	32.5
Processed Dairy Products	57	*	1.6
Apparel	56	*	.5
Sugar	42	*	3.6
Food Grains and Feed Crops	38	*	.3
Alcoholic Beverages	38	*	1.4
Canvas Products	29	*	29.9
All Other Secondary Triaggregate Sectors	76	*	*
Total	$2,308	.8%	2.1%†

* Less than 0.05%.

† All other Secondary Triaggregate sectors were each less than 0.1% of the inputs to this aggregate, or less than 0.1% of the outputs of this aggregate.

SALES TO THE SECONDARY TRIAGGREGATE SECTORS:

Secondary Triaggregate Sectors	Millions of Dollars	As % of Total Sales by the All Other Industries Aggregate	% of Total Input to Each Secondary Triaggregate Sector Procured from the All Other Industries Aggregate
Apparel	$297	.2%	2.6%
Food Grains and Feed Crops	209	.1	1.9
Spinning, Weaving, and Dyeing	141	.1	1.7
Animal Oils	120	.1	15.4
Miscellaneous Food Products	111	.1	1.7
Meat Animals and Products	98	.1	1.0
Leather and Leather Goods	97	.1	2.6
Grain Mill Products	89	*	1.7
Meat Packing and Wholesale Poultry	86	*	3.3
Vegetables and Fruits	82	*	2.0
All Other Agriculture	64	*	.8
Canning, Preserving, and Freezing	60	*	2.2
Farm Dairy Products	54	*	1.1
Alcoholic Beverages	42	*	1.5
House Furnishings and Other Nonapparel	42	*	2.3
Bakery Products	37	*	1.1
Fishing, Hunting, and Trapping	34	*	8.4
Cotton	30	*	1.4
Poultry and Eggs	28	*	.7
Processed Dairy Products	28	*	.8
Oil-Bearing Crops	19	*	1.8
Sugar	18	*	1.5
Vegetable Oils	18	*	.9
All Other Secondary Triaggregate Sectors	56	*	1.2
Total	$1,860	1.1%	1.7%†

* Less than 0.05%.

† Per cent of Secondary Triaggregate total outputs purchased by, or inputs procured from, All Other Industries Aggregate.

[39] Derived from Appendix Tables A-14 and A-33 and Exhibit 2.

EXHIBIT 12. PURCHASES BY AND SALES OF THE GOVERNMENT SERVICES AND GOVERNMENT PURCHASES SECTOR: 1947

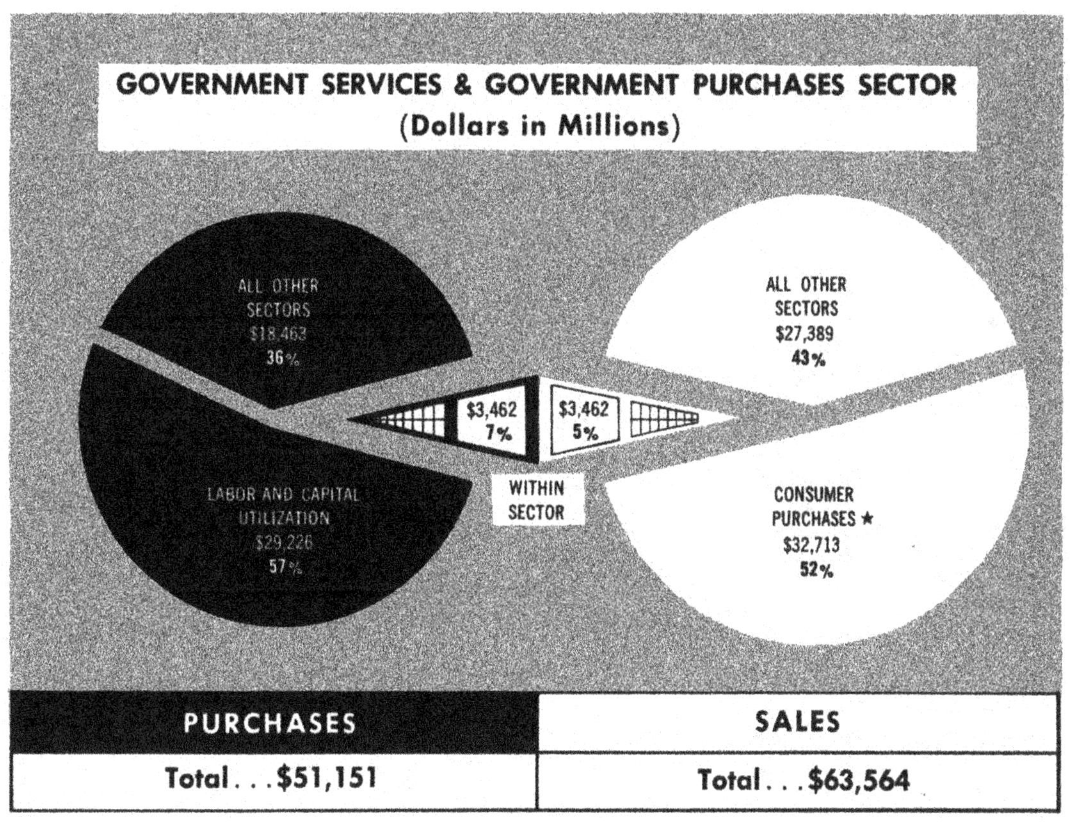

★ Plus Gross Private Capital Formation and Inventory Changes.

Government Sectors. In 1947 the cash outlays of the federal, state, and local governments totaled over $51 billion. The pie chart, Exhibit 12, indicates in general the relationship between purchases made by the government and the revenues it collected from the economy for services.

Government Purchases[45] is divided into three broad categories: 57% in the form of factor payments (primarily wages); 36% for goods and services from the Processing Aggregates; and the remaining 7% for goods and services within its own sector (primarily transfer payments among government agencies).[46]

Government Sales (Taxes) represents total revenues (tax and nontax receipts on current account)[47]

[45] The government columns show expenditures for goods and services, including purchases of capital goods and transfer payments. All new public construction and maintenance (including force account) were treated as purchases directly from the respective construction sectors rather than from the sectors providing construction ingredients; e.g., materials, services, wages, and salaries. Similarly, government expenditures for health and education were treated as purchases directly from the hospital and education industries rather than from sectors providing the items pertaining to such activities. However, purchases of equipment pertaining to government activities, such as that used in public construction and in operation of public hospitals and schools, were charged to the government account. Government interest payments (except payments to social insurance funds) and unilaterals were handled on a net basis. SOURCE: BLS Report No. 33—1947 Interindustry Relations Study, p. 33.

[46] Government payments of interest to social insurance funds and contributions to such funds were considered as real costs to government for services rendered and therefore were included in intragovernment transactions. They were considered to be wage supplements in the same sense as employer contributions to social insurance. The intragovernment transactions also included payments from one government sector to another, such as federal grants-in-aid to the states. SOURCE: BLS Report No. 33, p. 33.

[47] Corporate income tax receipts were estimated on an accrual basis; other tax receipts on a collection basis allocated to the industries legally liable for payment, except for the prorated tax liability resulting from interest income—which was allocated to Consumer Purchases (Households), since interest was treated in

(footnote continued on page 56)

TABLE 14. GOVERNMENT SECTOR PURCHASES: 1947

Sellers	Rank within Sector	Gross Purchases Dollars (Millions)	% of Total
Labor and Capital Utilization	1	$29,226	57.1%
Government Services (Taxes, etc.)	2	3,462	6.8
Imports (Noncompetitive)	3	2,345	4.6
Maintenance and Construction	4	1,936	3.8
Railroads	5	285	0.5
Meat Packing and Wholesale Poultry	6	260	0.5
Petroleum Products	7	229	0.4
Water and Other Transportation	8	216	0.4
Real Estate and Rentals	9	198	0.4
Spinning, Weaving, and Dyeing	10	148	0.3
Apparel	11	142	0.3
Electric Light and Power	12	139	0.3
Other Chemical Industries	13	126	0.2
Processed Dairy Products	14	115	0.2
Vegetables and Fruits	15	90	0.2
Trucking	16	80	0.2
Miscellaneous Food Products	17	69	0.1
Other Fuels (Coke, Coal, and Gas)	18	68	0.1
Paper and Board Mills and Converted Paper Products	19	64	0.1
Automobile and Other Repair Services	20	57	0.1
Poultry and Eggs	21	56	0.1
Canning, Preserving, and Freezing	22	55	0.1
Sugar	23	46	0.1
Leather and Leather Goods	23	46	0.1
Grain Mill Products	25	44	0.1
Wholesale Trade	26	37	0.1
Jute, Linens, Cord, and Twine	27	31	0.1
Tires, Tubes, and Other Rubber Products	28	29	0.1
Farm Dairy Products	29	26	0.1
Canvas Products	30	25	*
Special Textile Products	31	23	*
Metal Container Materials and Cork Products	32	17	*
Advertising	33	13	*
House Furnishings and Other Nonapparel	34	11	*
Vegetable Oils	35	9	*
Meat Animals and Products	36	7	*
Bakery Goods	37	6	*
Wood Containers and Cooperage	38	4	*
Fishing, Hunting, and Trapping	39	2	*
Glass	39	2	*
Fertilizers	39	2	*
Warehouse and Other Transportation	39	2	*
Food Grains and Feed Crops		†	**
All Other Agriculture		†	**
Alcoholic Beverages		†	**
Animal Oils		†	**
Tin Cans and Other Tins		†	**
Retail Trade		†	**
All Other Industries	—	11,403	22.3
Total	—	$51,151	100.0%

* Less than 0.05%.
** Less than 0.05%; note that these amounts are not included in the total.
† Sales of less than $500,000; note that these amounts are not included in the total.
SOURCE: Derived from data published by the Bureau of Labor Statistics, Division of Interindustry Economics.

TABLE 15. GOVERNMENT SERVICES (TAXES, ETC.) SECTOR SALES: 1947

Buyers	*Rank within Sector*	*Gross Sales* Dollars (Millions)	*% of Total*
Consumer Purchases (Households)	1	$32,432	51.0%
Government Purchases	2	3,462	5.4
Eating and Drinking Places	3	1,413	2.2
Exports (minus Competitive Imports)	4	1,251	2.0
Spinning, Weaving, and Dyeing	5	549	0.9
Food Grains and Feed Crops	6	348	0.5
Apparel	7	342	0.5
Miscellaneous Food Products	8	317	0.5
Saw Mills, Planing, and Veneer Mills	9	225	0.4
Gross Private Capital Formation	10	216	0.3
Alcoholic Beverages	11	176	0.3
Medical, Dental, and Other Professional Services	12	167	0.3
Leather and Leather Goods	13	137	0.2
Meat Animals and Products	14	134	0.2
Tires, Tubes, and Other Rubber Products	15	120	0.2
Grain Mill Products	16	113	0.2
Meat Packing and Wholesale Poultry	17	112	0.2
Canning, Preserving, and Freezing	18	111	0.2
Bakery Products	19	102	0.2
Vegetables and Fruits	20	98	0.2
Processed Dairy Products	21	94	0.1
Vegetable Oils	22	93	0.1
Tobacco Manufactures	23	81	0.1
Paints and Allied Products	24	70	0.1
Soap and Related Products	25	69	0.1
Farm Dairy Products	26	67	0.1
Inventory Changes	27	65	0.1
Cotton	28	58	0.1
All Other Agriculture	29	55	0.1
House Furnishings and Other Nonapparel	30	52	0.1
Sugar	31	46	0.1
Special Textile Products	31	46	0.1
Poultry and Eggs	33	38	0.1
Oil-Bearing Crops	34	20	*
Animal Oils	35	19	*
Tobacco	36	12	*
Fishing, Hunting, and Trapping	36	12	*
Jute, Linens, Cord, and Twine	38	11	*
Gum and Wood Chemicals	39	7	*
Canvas Products	40	3	*
All Other Industries	—	20,821	32.7
Total	—	$63,564	100.0%

* Less than 0.05%.

SOURCE: Derived from data published by the Bureau of Labor Statistics, Division of Interindustry Economics.

for services rendered to the economy. The allocation of these receipts in the economy was as follows: 52% was obtained from personal income taxes (end-product consumers); 43% from corporate taxes (processing or intermediate sectors); and 5% from intragovernment activity.[48] (*See p. 56.*)

Tables 14 and 15 indicate more completely the government purchases (inputs) and the sources of government taxes for services (outputs). Relating the Secondary Triaggregate to these government sectors, one obtains the following abbreviated tables:

Transactions Between Government and the Secondary Triaggregate,[49] *1947*

GOVERNMENT PURCHASES FROM THE SECONDARY TRIAGGREGATE SECTORS:

Secondary Triaggregate Sectors	*Millions of Dollars*	*As % of Total Purchases by Government Sector*	*% of Total Output of Each Secondary Triaggregate Sector Purchased by the Government*
Meat Packing, and Wholesale Poultry	$260	.5%	2.3%
Spinning, Weaving, and Dyeing	148	.3	1.8
Apparel	142	.3	1.3
Processed Dairy Products	115	.2	3.2
Vegetables and Fruits	90	.2	2.2
Miscellaneous Food Products	69	.1	1.0
Poultry and Eggs	56	.1	1.4
Canning, Preserving, and Freezing	55	.1	2.0
Sugar	46	.1	3.9
Leather and Leather Products	46	.1	1.2
Grain Mill Products	44	.1	.8
Jute, Linens, Cord, and Twine	31	.1	12.2
Farm Dairy Products	26	.1	.5
Canvas Products	25	*	25.8
Special Textile Products	23	*	2.8
All Other Triaggregate Sectors	35	*	*
Totals	$1,211	2.4%	1.1%†

* Less than 0.05%.
† Percentage of total output of the Secondary Triaggregate purchased by the government, and percentage of total input of the Secondary Triaggregate purchased from the government.

GOVERNMENT SALES (TAXES) TO THE SECONDARY TRIAGGREGATE SECTORS:[50]

Secondary Triaggregate Sectors	*Millions of Dollars*	*As % of Total Taxes Paid to the Government*	*% of Total Inputs of Each Secondary Triaggregate Sector Paid to the Government*
Spinning, Weaving, and Dyeing	$549	.9%	6.8%
Food Grains and Feed Crops	348	.5	3.2
Apparel	342	.5	3.0
Miscellaneous Food Products	317	.5	4.8
Alcoholic Beverages	176	.3	6.5
Leather and Leather Goods	137	.2	3.7
Meat Animals and Products	134	.2	1.4
Grain Mill Products	113	.2	2.1
Meat Packing and Wholesale Poultry	112	.2	1.0
Canning, Preserving, and Freezing	111	.2	4.1
Bakery Products	102	.2	3.0
Vegetables and Fruits	98	.2	2.4
Processed Dairy Products	94	.1	2.6
Vegetable Oils	93	.1	4.9
Tobacco Manufactures	81	.1	3.2
Farm Dairy Products	67	.1	1.3
Cotton	58	.1	2.6
All Other Agriculture	55	.1	2.8
House Furnishings and Other Nonapparel	52	.1	2.9
Sugar	46	.1	3.9
Special Textile Products	46	.1	5.6
Poultry and Eggs	38	.1	1.0
All Other Secondary Triaggregate Sectors	77	*	2.2
Totals	$3,246	5.2%	3.0%†

The above transactions are relatively small as a percentage of the total inputs (purchases) or taxes (outputs) of the Government sectors or of the combined Secondary Triaggregate. Nevertheless, they are important to the respective sectors of the Secondary Triaggregate as the totals of $3.2 bil-

(*footnote 47 continued*)
the present study as an output of Labor and Capital Utilization. Excise taxes, including general sales taxes, were handled as margin items; i.e., they were allocated to industries purchasing products or services upon which these taxes applied, rather than to industries legally responsible for payment. Customs duties associated with competitive imports were allocated to the comparable domestic producing industry while customs duties associated with noncompetitive imports were allocated to the using industries. SOURCE: BLS Report No. 33, p. 33.

[48] This intragovernment activity as a percentage of output (revenues) is smaller because total government revenues were greater than purchases in 1947. The dollar amount of intragovernment activity was the same. See footnote 46, page 53.

[49] Derived from Tables 14 and 15 and Exhibit 2.

[50] In this context sales receipts take the form of tax receipts.

lion taxes and $1.2 billion government purchases indicate.

Next we shall analyze the purchases by the private end-product consuming sectors in the economy and compare these with the labor and capital costs used to produce the items that consumers purchased.

Consumer Purchases and Labor and Capital Utilization. The Consumer Purchases (Household) sector on the output side of the matrix, consisting primarily of expenditures for personal consumption,[51] is partially offset on the input side by the Labor and Capital Utilization sector, representing costs of labor and capital to produce items for final consumption.[52] These two sectors do not offset each other but are used here to illustrate the relative importance of the Consumer Purchases column (column 44, Exhibit 2) (household column in Leontief system) to the processing sector of the agribusiness economy and the relative importance of Labor and Capital Utilization (row 58, Exhibit 2) (household row in Leontief system) to the agribusiness sector of the economy.

Tables 16 and 17 present in more detail the purchases by consumers (Consumer Purchases) and the distribution of labor and capital costs (Labor and Capital Utilization) in the production of the goods and services involved. The general relationship of these sectors to the Secondary Triaggregate of agribusiness can be ascertained from the tabulations shown on page 60.

These tabulations indicate not only the general importance of Consumer Purchases to the Secondary Triaggregate, but also its substantial relationship to each of the industry sectors within the Secondary Triaggregate. Similarly, the reader is able to ascertain the considerable importance of labor, depreciation, and the other factor and nonfactor payment items to the sectors comprising the Secondary Triaggregate.

Let us now turn to the final section of this chapter and consider the direct and indirect flow of certain goods and services in the economy that can be measured quantitatively through the use of the input-output technique.

[51] Consumer Purchases ("Household" column) is comprised primarily of personal consumption expenditures, including those of farm households for personal living requirements, direct personal taxes, and imputed charges for food produced and consumed on farms. (Costs of farming operations are excluded.) Householders' purchases of dwelling units for their own occupancy were not included, such transactions being treated as a business investment and allocated to Gross Private Capital Formation. Some expenditures by individuals in connection with their business activities were included, such as hand tools purchased by carpenters. Expenses of individuals for travel related to their business activities were for the most part included. Since the individual cell entries in the matrix were expressed in producers' values, such subsequent charges as transportation costs, trade margins, and excise taxes relating to household purchases of goods and services were shown as direct payments by households to the sectors incurring these distributive cost items. Sales taxes were treated in the same manner as excise taxes, except that those reported as part of trade margins were treated as tax payments and covered in household payments to trading firms. Secondhand household purchases were considered only to the extent of the gross trade margins on such items.

For consistency imputed rentals were treated as actual—all rental payments covering utilities (heat, light, etc.). This differs from the space rent concept used by the Department of Commerce in its consumption expenditures series in that the latter excluded all costs for utilities. In general, maintenance of residential buildings was charged as a cost to the rental industry rather than to households. (A small outlay for maintenance represented actual maintenance outlays by tenants not appearing as costs of the rental industry.)

[52] Labor and Capital Utilization ("Household" row) represents essentially all charges against final demand, except payments to foreign trade (noncompetitive imports) and payments to government (all taxes, including income taxes). These items may be segregated into factor and nonfactor charges; the former consisting of wages and salaries, employer contribution to private pension plans, royalties, interest, entrepreneurial income, and corporate profits (after taxes), and the latter of transfer payments (including contributions and gifts), depreciation and amortization, capital outlays charged to current expense, losses, and accidental damage to fixed capital (uninsured), business travel and entertainment (including reimbursement for personal car use), banking service cash charges to business, and claim payments (primarily nonlife insurance). Since essentially these were considered as cost items, not treated as payments to individuals, they were included in the Labor and Capital Utilization row for a variety of reasons. Capital consumption charges which might properly belong in a gross private capital formation row, were included in households because of the difficulties of segregating these charges for each individual industry from the total charges against the final product of that industry. Business travel and entertainment charges were included here to offset allocations to households on the product side, which included purchases by individuals of transportation and entertainment for business uses. Banking service cash charges to business were included here to offset the allocation to households of the portion of banking output (services) applicable to business. Transfer payments, included in Labor and Capital Utilization, as noted earlier, were part of the Labor and Capital Utilization entry for the industry involved in such payments. Thus, government transfer payments are part of the Labor and Capital Utilization entries in the government columns. This means that the profits of the receiving industries had to be adjusted to exclude subsidies—otherwise, the outlays of those industries would have exceeded, by corresponding amounts, the revenue derived from the sales.

The payments represented by Labor and Capital Utilization correspond, in the main, to national income—adjusted to exclude employers' payments of payroll taxes and corporate income taxes, and to include capital consumption allowances, individuals' receipts of insurance claims and bad debt allowances. SOURCE: BLS Report No. 33—1947 Interindustry Relations Study.

TABLE 16. CONSUMER PURCHASES (HOUSEHOLDS) SECTOR: 1947

Sellers	*Rank within Sector*	*Gross Purchases* Dollars (Millions)	% of Total
Government Services (Taxes, etc.)	1	$32,432	16.4%
Retail Trade	2	21,510	10.9
Real Estate and Rentals	3	20,289	10.3
Apparel	4	8,982	4.5
Meat Packing and Wholesale Poultry	5	8,322	4.2
Wholesale Trade	6	6,224	3.2
Miscellaneous Food Products	7	4,053	2.1
Bakery Products	8	3,013	1.5
Farm Dairy Products	9	2,731	1.4
Railroads	10	2,680	1.4
Vegetables and Fruits	11	2,651	1.3
Poultry and Eggs	12	2,589	1.3
Automobile and Other Repair Services	13	2,585	1.3
Petroleum Products	14	2,440	1.2
Labor and Capital Utilization	15	2,116	1.1
Processed Dairy Products	16	2,086	1.1
Leather and Leather Goods	17	2,065	1.0
Canning, Preserving, and Freezing	18	2,064	1.0
Other Chemical Industries	19	1,988	1.0
Tobacco Manufactures	20	1,485	0.8
Alcoholic Beverages	21	1,209	0.6
Meat Animals and Products	22	1,070	0.5
Grain Mill Products	23	993	0.5
House Furnishings and Other Nonapparel	24	957	0.5
Trucking	25	925	0.5
Spinning, Weaving, and Dyeing	26	919	0.5
Water and Other Transportation	27	890	0.5
Imports (Noncompetitive)	28	833	0.4
Tires, Tubes, and Other Rubber Products	29	731	0.4
All Other Agriculture	30	641	0.3
Sugar	31	484	0.2
Special Textile Products	32	451	0.2
Paper and Board Mills and Converted Paper Products	33	365	0.2
Warehouse and Storage	34	215	0.1
Metal Container Materials and Cork Products	35	187	0.1
Maintenance and Construction	36	154	0.1
Glass	37	148	0.1
Electric Light and Power	38	133	0.1
Fishing, Hunting, and Trapping	39	85	*
Advertising	40	49	*
Food Grains and Feed Crops	41	45	*
Jute, Linens, Cord, and Twine	42	26	*
Canvas Products	43	25	*
Vegetable Oils	44	19	*
Tin Cans and Other Tins	45	18	*
Fertilizers	46	8	*
Wood Containers and Cooperage	47	7	*
Oil-Bearing Crops	48	3	*
Other Fuels (Coal, Coke, and Gas)	49	3	*
All Other Industries	—	53,639	27.1
Total	—	$197,537	100.0%

* Less than 0.05%.

SOURCE: Derived from data published by the Bureau of Labor Statistics, Division of Interindustry Economics.

TABLE 17. LABOR AND CAPITAL UTILIZATION SECTOR SALES: 1947

Buyers	*Rank within Sector*	*Gross Sales* *Dollars (Millions)*	*% of Total*
Government Purchases	1	$29,226	12.8%
Medical, Dental, and Other Professional Services	2	6,832	3.0
Food Grains and Feed Crops	3	6,273	2.7
Eating and Drinking Places	4	4,625	2.0
Apparel	5	4,609	2.0
Meat Animals and Products	6	3,270	1.4
Spinning, Weaving, and Dyeing	7	3,025	1.3
Vegetables and Fruits	8	2,869	1.2
Consumer Purchases (Households)	9	2,116	0.9
Farm Dairy Products	10	1,811	0.8
Saw Mills, Planing, and Veneer Mills	11	1,582	0.7
Meat Packing and Wholesale Poultry	12	1,579	0.7
Miscellaneous Food Products	13	1,572	0.7
Cotton	14	1,501	0.7
All Other Agriculture	15	1,386	0.6
Leather and Leather Goods	16	1,288	0.6
Bakery Products	17	1,248	0.5
Tires, Tubes, and Other Rubber Products	18	1,168	0.5
Alcoholic Beverages	19	1,073	0.5
Exports (minus Competitive Imports)	20	914	0.4
Grain Mill Products	21	796	0.3
Canning, Preserving, and Freezing	22	687	0.3
Tobacco	23	656	0.3
Oil-Bearing Crops	24	621	0.3
Processed Dairy Products	25	615	0.3
Paints and Allied Products	26	521	0.2
Tobacco Manufactures	27	477	0.2
House Furnishings and Other Nonapparel	28	397	0.2
Soap and Related Products	29	388	0.2
Special Textile Products	30	356	0.2
Poultry and Eggs	31	323	0.1
Fishing, Hunting, and Trapping	32	301	0.1
Vegetable Oils	33	243	0.1
Gross Private Capital Formation	34	232	0.1
Sugar	35	160	0.1
Animal Oils	36	126	0.1
Jute, Linens, Cord, and Twine	37	81	*
Gum and Wood Chemicals	38	66	*
Canvas Products	39	34	*
All Other Industries	—	143,849	62.8
Total	—	$228,896	100.0%

* Less than 0.05%.

SOURCE: Derived from data published by the Bureau of Labor Statistics, Division of Interindustry Economics.

Relationship of Consumer Purchases from and Factor Payments to the Secondary Triaggregate, 1947[53a]

CONSUMER PURCHASES (HOUSEHOLDS)[53] (COLUMN 44) FROM THE SECONDARY TRIAGGREGATE SECTORS:

Secondary Triaggregate Sectors	*Millions of Dollars*	*As % of All Purchases by Consumer Purchases Sector*	*% of Total Output of Each Secondary Triaggregate Sector Purchased by Consumers*
Apparel	$8,982	4.5%	79.2%
Meat Packing and Wholesale Poultry	8,322	4.2	74.3
Miscellaneous Food Products	4,053	2.1	61.1
Bakery Products	3,013	1.5	89.9
Farm Dairy Products	2,731	1.4	54.0
Vegetables and Fruits	2,651	1.3	66.1
Poultry and Eggs	2,589	1.3	67.0
Processed Dairy Products	2,086	1.1	57.2
Leather and Leather Goods	2,065	1.0	55.3
Canning, Preserving, and Freezing	2,064	1.0	75.7
Tobacco Manufactures	1,485	.8	57.8
Alcoholic Beverages	1,209	.6	44.4
Meat Animals and Products	1,070	.5	10.9
Grain Mill Products	993	.5	18.6
House Furnishings and Other Nonapparel	957	.5	53.0
Spinning, Weaving, and Dyeing	919	.5	11.4
All Other Agriculture	641	.3	32.8
Sugar	484	.2	41.0
Special Textile Products	451	.2	54.9
Fishing, Hunting, and Trapping	85	*	21.0
Food Grains and Feed Crops	45	*	.4
Jute, Linens, Cord, and Twine	26	*	10.2
Canvas Products	25	*	25.8
Vegetable Oils	19	*	1.0
Oil-Bearing Crops	3	*	.3
Totals	$46,968	23.6%	43.3%†

* Less than 0.05%.
† Per cent of total Secondary Triaggregate outputs purchased by consumers.
[53] Another illustration of consumer purchases from agribusiness industry sectors is found in the Flow Chart (Exhibit 3). The reader must keep in mind that in the Input-Output Chart, Flow Chart, pie charts, and tables, the amounts are in producers' values—trade margins, transportation, etc., being included in the Input-Output Chart as separate sectors. Consumer Purchases from agribusiness as presented in the Flow Chart represents 43% of total Consumer Purchases.

DISTRIBUTION OF LABOR AND CAPITAL UTILIZATION (ROW 58) OUTPUTS AMONG SECONDARY TRIAGGREGATE SECTORS:

Secondary Triaggregate Sectors	*Millions of Dollars*	*As % of Total Output of Labor and Capital Utilization Sector*	*% of Total Input of Each Secondary Triaggregate Sector Procured from Labor and Capital Utilization Sector*
Food Grains and Feed Crops	$6,273	2.7%	57.0%
Apparel	4,609	2.0	40.7
Meat Animals and Products	3,270	1.4	33.4
Spinning, Weaving, and Dyeing	3,025	1.3	37.4
Vegetables and Fruits	2,869	1.2	71.5
Farm Dairy Products	1,811	.8	35.7
Meat Packing and Wholesale Poultry	1,579	.7	14.2
Miscellaneous Food Products	1,572	.7	23.7
Cotton	1,501	.7	67.6
All Other Agriculture	1,386	.6	70.8
Leather and Leather Goods	1,288	.6	34.5
Bakery Products	1,248	.5	37.2
Alcoholic Beverages	1,073	.5	39.4
Grain Mill Products	796	.3	14.9
Canning, Preserving, and Freezing	687	.3	25.2
Tobacco	656	.3	74.1
Oil-Bearing Crops	621	.3	58.6
Processed Dairy Products	615	.3	16.9
Tobacco Manufactures	477	.2	18.6
House Furnishings and Other Nonapparel	397	.2	22.0
Special Textile Products	356	.2	43.3
Poultry and Eggs	323	.1	8.4
Fishing, Hunting, and Trapping	301	.1	74.3
Vegetable Oils	243	.1	12.7
Sugar	160	.1	13.6
Animal Oils	126	.1	16.2
Jute, Linens, Cord, and Twine	81	*	32.0
Canvas Products	34	*	35.1
Totals	$37,377	16.4%	34.5%†

* Less than 0.05%.
† Per cent of total inputs of the Secondary Triaggregate purchased from Labor and Capital Utilization.
[53a] Derived from Tables 16 and 17 and Exhibit 2.

DIRECT AND INDIRECT AGRIBUSINESS RELATIONSHIPS

The purpose of this section is to indicate additional uses of input-output methodology in analyzing not only direct or primary agribusiness relationships but also certain indirect or secondary effects on the rest of the economy.

In doing this, the authors make use of the 50-industry input-output charts, derived by Evans and Hoffenberg (Exhibits 13, 14, and 15), which divide the economy into 45 processing and 5 final demand and final payment sectors. (These three exhibits will be found at the back of the book and may be unfolded for easy reference.) Since Evans and Hoffenberg derived their charts from B.L.S. data, they conform, generally, to the sector arrangement used in the B.L.S. Interindustry Study and the Harvard Economic Research Project.

METHODOLOGY

The Interindustry Flow of Goods and Services by Industry of Origin and Destination, 1947 (Exhibit 13) may be reconciled with Agribusiness and the National Economy Input-Output Chart (Exhibit 2)[54] if one keeps the following data in mind:

The Agriculture and Fisheries Sector of Exhibit 13 corresponds to the Farming Aggregate of Exhibit 2. The difference in total gross output occurred in the process of refining the 1947 census data and reallocating by-products among the original producers of such items.

Inventory Changes (additions are in columns and depletions in rows) in Exhibit 13 were combined as net additions or depletions in the Inventory Changes column in Exhibit 2.

Household, as a column in Exhibit 13, corresponds to the Consumer Purchases column in Exhibit 2—the rows, Households and Gross Private Capital Formation, in Exhibit 13 correspond to the row, Labor and Capital Utilization in Exhibit 2.

Foreign Countries (exports to) as a column and Foreign Countries (imports from) as a row in Exhibit 13 correspond to Exports (minus Competitive Imports) as a column in Exhibit 2.

One may use Exhibit 13 in the same manner as Exhibit 2, recognizing that Exhibit 13 represents the economy in balanced industry sectors, whereas Exhibit 2, being agribusiness oriented, combines much of the economy in one aggregate. Row 1, column 1, in Exhibit 13 indicates that the Agriculture and Fisheries sector, itself, consumed $10,856 million worth of the total goods it produced. Continuing along row 1, one notes that Food and Kindred Products (column 2) purchased $15,048 million of goods and services from Agriculture and Fisheries; Tobacco Manufactures (column 3) purchased $783 million, etc.

Exhibit 14, Direct Purchases per Dollar of Output, was computed from Exhibit 13 and contains the unit cost structure for each of the 45 processing sectors set forth in Exhibit 13. This unit cost structure is stated for each sector in terms of its direct purchases from other industries. The columns of this input-output chart represent purchases, as in Exhibits 13 and 2, except that direct purchases from a producing industry is presented in terms of the ratio of each such purchase to the total outlays of the purchasing industry. For example, reading from top to bottom in column 1 (Exhibit 14), each dollar of current output from Agriculture and Fisheries required intrasector transfers of 26.1 cents (for such items as feeder cattle, feed, seed, etc.); purchases from the Food and Kindred Products sector of 5.7 cents (for prepared feeds, etc.); and so on *ad infinitum*. Each of the other columns may be read in a similar manner.[55]

In the preparation of Exhibit 14 the ratios were computed in two stages.[56] First, the gross output figure for each sector shown in Exhibit 13 was adjusted to represent net output during 1947 by subtracting inventory depletion. To illustrate, gross output for Agriculture and Fisheries (sector 1) amounting to $44,263 million was reduced by $2,660 million (row 46, column 1) to give an adjusted gross output figure of $41,603 million for this sector.[57] Secondly, this adjusted gross output for 1947 was divided into each of the sectors of the Agriculture and Fisheries column (working from top to bottom, sector by sector, of

[54] The only differences are those of aggregation and nomenclature with the exception of the refinement of gross output and outlays that occurred after 1952 (the date of publication of these three charts).

[55] The reader is cautioned that the ratios reflect only the cost and price structure prevailing in 1947.

[56] "The Interindustry Relations Study for 1947," pp. 97–142.

[57] Ibid., p. 138.

column 1, Exhibit 13) giving the ratios shown in Exhibit 14. To illustrate, the intrasector transactions within the Agriculture and Fisheries sector (row 1, column 1, Exhibit 13) of $10,856 million (for such items as feeder cattle, feed, and seed), divided by $41,603 million gives a quotient of 26.1 cents—the entry found in column 1, row 1, Exhibit 14. In a like manner the purchases of Agriculture and Fisheries from the Food and Kindred Products sector of $2,378 million (column 1, row 2, Exhibit 13), divided by $41,603 million, yields 5.7 cents (column 1, row 2, Exhibit 14).[58] Similarly derived were the ratios for the other sectors of Exhibit 14. These ratios would add up to $1.00 for the columns of each sector if factor payment ratios were included in Exhibit 14.[59]

The utility of Exhibit 14 is that from it one may determine the value of inputs (costs) from processing sectors required by a given sector to produce $1.00 of output. For example, following down column 1 (Agriculture and Fisheries) one notes that Agriculture requires 26.1 cents of direct inputs from itself to produce $1 of output; 5.7 cents from Food and Kindred Products; 0.15 cents from Textile Mill Products, etc.[60]

Exhibit 15 represents both the direct and indirect requirements of the various sectors of the economy needed to produce $1 of product in terms of purchases by the final consumer. Just as a stone thrown in a pool makes ripples in ever-widening circles, so do the actions of each industry affect other areas of the economy. This analogy holds even to the point that when the "ripples" hit the "shore line" they bounce back against the original ripples, forming an intertwining effect. In a general way this interaction between ripples illustrates the direct and indirect effects of activity in one sector of the economy on other sectors.

Exhibit 15 shows the combined direct and indirect requirements placed on all sectors of the economy by the delivery of $1 of output outside of the processing system. In interindustry terminology the phrase "Outside of the processing system" means delivery to sectors which do not perform a processing function, i.e., consumers, government, exports, inventories, etc. In business terminology Exhibit 15 sets forth the inputs (costs) required by a given sector in producing and selling a product to the final consumer, as well as the inputs (costs) of the sectors that supplied the given sector. For example, a soybean processing firm has to pay farmers for the soybeans it uses (a direct input) and the farmer who grew the soybeans in turn purchased fertilizer, gasoline, etc. These direct costs (inputs) to the farmer constitute indirect costs (inputs) to the soybean processor. In this manner every industry sector directly or indirectly supports every other sector.

In using Exhibit 15 the reader should note that the rows and columns have been reversed as compared to Exhibits 13 and 14. Thus, if we look at Exhibit 15, row 1 is comparable to column 1 of Exhibits 13 and 14. The most frequent use of Exhibit 15 is to read across the horizontal rows in which the data indicate the direct and indirect requirements from the industry sectors named at the top to produce $1 of final demand by the industry sectors named at the left. For example, row 1 of Exhibit 15 indicates the requirements from each of the sectors of the economy to produce $1 of final demand by Agriculture and Fisheries. We may interpret the vertical columns in Exhibit 15 as we did the rows in Exhibits 13 and 14. The vertical columns in Exhibit 15 indicate the relative importance of the sector named at the top in supplying the requirements per dollar of final demand by sectors named at the left. For example, column 1 of Exhibit 15 indicates the relative importance of Agriculture and Fisheries to the production of a dollar of final demand by each industry sector at the left.

The value of Exhibit 15 is that it provides a basis for computing the direct and indirect effect per sector of a given increase or decrease in final demand. If the data presented in Exhibit 15 were in sufficient detail, one might obtain from it complete marketing and purchasing information for a given industry. For example, one might trace back from end products the direct and indirect demand on the steel industry. Thus quantitative relation-

[58] It should be noted that these ratios are in terms of 1947 data and would be different for other years.

[59] Ratios have been computed only for processing sectors.

[60] The reader will notice that row 20, column 1 shows only a small amount of Farm Machinery as an input to Agriculture—as previously stated, see footnote 20, page 28. This item is considered part of Gross Private Capital Formation (or depreciation) but not as a direct input of the Agriculture and Fisheries sector.

ships between production and ultimate demand can be traced through the economy for this or any other sector.

Exhibit 15 has two sets of data: (1) Values appearing along the major diagonal—i.e., column 1, row 1; column 2, row 2; column 3, row 3, etc.—represent requirements of a specific sector to produce a dollar of goods and services for the consumer from such sector. These values also include (a) the direct requirements of that sector, in the form of intrasector transactions in satisfying $1.00 of final demand, and (b) the indirect requirements of such sector from other sectors, in satisfying such final demand. (2) All the other values shown in Exhibit 15 indicate (a) the direct requirements of any given sector from all other sectors in producing $1.00 of final demand, and (b) the indirect requirements of such sector from other sectors in satisfying such demand.

Exhibit 15 is derived from Exhibit 14 by means of a series of related calculations which trace both the direct and indirect effects of the purchase of $1.00 of end products at the consumer level.

To illustrate the method of computation, let us take as an example the value shown in column 1, row 1 of Exhibit 15. The $1.41445 item shown there represents the sum of the original dollar of demand plus the direct and indirect requirements from the Agriculture and Fisheries sector to produce $1.00 of final demand from the Agriculture and Fisheries sector. The computation of the $1.41445 value in Exhibit 15 is as follows: One dollar of final demand is assumed to have flowed from the Agriculture and Fisheries sector to the ultimate consumer. To this is added the $.260935 required in the form of direct intrasector transactions within Agriculture and Fisheries (column 1, row 1, Exhibit 14), making a total of at least $1.260935 required of the Agriculture and Fisheries sector for it to deliver $1.00 of output "outside the processing system." This $1.260395, in turn, will induce other indirect requirements from the given industry. The values for the supporting industries, which are shown in the Agriculture and Fisheries column of Exhibit 14, were computed on the basis of $1.00 of output and did not include the impact of the additional $.260935 of intrasector transactions required within Agriculture and Fisheries. This additional impact gives rise to further requirements in other sectors supplying Agriculture and Fisheries, and because of the interdependence of all sectors, the change in any one will have a further effect on the others. The aggregate of these effects will, in turn, require additional output of the Agriculture and Fisheries sector, and this additional output, in turn, will initiate further reciprocating transactions among all sectors on an ever-diminishing scale, *ad infinitum*. The sum of these indirect effects (on the Agriculture and Fisheries sector) generated by the impact of the additional direct intrasector requirement of $.260935 amounts to $.153515, which, when added to the original $1.00 plus the $.260935, gives a total of $1.41445—the amount that appears in Exhibit 15, column 1, row 1.

An example of a first-round effect of the direct and indirect requirements per $1.00 of final demand is as follows: the Food and Kindred Products relationship to Agriculture and Fisheries is $1.260935 x $.057168[61] x $.404140[62] = $.029132; which is the additional amount required of Agriculture and Fisheries by Food and Kindred Products for the latter to feed back to Agriculture and Fisheries sufficient inputs to enable Agriculture and Fisheries to increase its output by $1.260935. Similar calculations are required for each other sector. However, the sum of these values will still be somewhat short of the value of $1.41445 shown for Agriculture and Fisheries in Exhibit 15 because it takes into account only the first round of indirect effects.[63] When the second, third, and fourth rounds, etc., are taken into account, the sum of the values will equal $1.41445.

A second illustration of the relationship of Exhibit 14 to Exhibit 15 may be seen by a comparison of the value $.057168 (column 1, row 2, Exhibit 14), required directly from Food and Kindred Products per dollar of output by Agriculture and Fisheries, to the $.09706 (row 1, column 2, Exhibit 15) value required both directly and indirectly from Food and Kindred Products per dollar of final demand from Agriculture and Fisheries.

61 Column 1, row 2, Exhibit 14.

62 Column 2, row 1, Exhibit 14.

63 Exhibit 15 has only 44 computations for each step since the 45th sector in Exhibit 14, "New Construction and Maintenance," was excluded because it is a capital sector rather than a processing or flow sector.

(It should be remembered that these values are corresponding amounts because a column in Exhibit 14 corresponds to a row in Exhibit 15.) The first step is to take into account the direct impact of the $1.260935 intrasector activity within the Agriculture and Fisheries sector upon the requirements from the Food and Kindred Products sector. The mechanics of calculating this first step would be as follows: $.057168 x $1.260935 = $.072085. This $.072085 of output by Food and Kindred Products, in turn, generates more output by Food and Kindred Products because of its intrasector requirements of $.13187[64] per dollar of output. The computation of this additional requirement is $.072085 x $.131870 = $.009506, which, when added to the $.072085, gives a total of $.081591. This procedure is continued *ad infinitum.*

In addition, there would be second, third, and fourth-round effects, etc. each of decreasing value. The computation of these successive indirect values would follow the same technique as in either of the above two examples, *ad infinitum.* The reader will notice that Exhibit 15, in contrast to Exhibit 14, has all spaces filled, thus illustrating that each industry of the economy is connected with every other, either directly, indirectly, or both.[65]

The result is that the amounts shown in each row in Exhibit 15 indicate the total gross output of the sectors shown at the top required to support the delivery "outside the processing system" of $1.00 of product listed at the left-hand margin. To illustrate, in 1947 the delivery of $1.00 by the Food and Kindred Products sector outside the processing system would have required $0.68 from the Agriculture and Fisheries sector (row 2, column 1); $1.20 from itself (row 2, column 2); $0.09 from the Chemical sector (row 2, column 10); etc.

Utilization

Having reviewed the derivation and content of Exhibits 13, 14, and 15, let us now examine some of the uses to which they can be put.[66] From Exhibit 13 one may determine the dollar amounts that industries purchase from any given sector. This is simply reading down the column. In addition, from the same exhibit it is possible to ascertain the major markets of each of the sectors by reading across a row. Table 18 lists at the left those sectors which are the major direct purchasers of the Agriculture and Fisheries sector (values obtained from row 1 in Exhibit 13). Exhibit 13 also indicates the direct markets for those sectors that purchase from Agriculture and Fisheries. The markets of the purchasers of Agriculture and Fisheries are listed in the third column of Table 18, and, in a sense, are a derived demand for the ultimate purchase of the products of the Agriculture and Fisheries sector. In Exhibit 13, we are unable to measure the magnitude of the effect of this derived or indirect demand instituted by the principal markets of the direct buyers from Agriculture and Fisheries. However, in Exhibit 15, such an analysis is possible.

Table 19 indicates the direct and indirect markets of Agriculture and Fisheries obtained from column 1, Exhibit 15. (As noted previously, in Exhibit 15 the rows and columns are reversed, hence column 1 in Exhibit 15 corresponds to row 1 in Exhibits 13 and 14.) In essence, Table 19 shows the exact same values as those in Exhibit 15 times one million. These amounts indicate the direct and indirect inputs required from Agriculture and Fisheries to produce $1 million of final demand in terms of those products requiring the most inputs from the Agriculture and Fisheries sector per dollar of output in 1947. Because indirect as well as direct requirements are shown in Exhibit 15, one is able to analyze the relative importance of the markets of the principal purchasers from Agriculture and Fisheries to Agriculture and Fisheries. As indicated in Table 19, certain derived demand markets, such as Tobacco Manufactures and Eating and Drinking Places, were more important to Agriculture and Fisheries, per dollar of their output, than was shown by just the direct purchases or direct demand indicated in Table 18. Table 19 indicates not only the importance of the requirements of the other sectors of the economy with respect to Agriculture and Fisheries, but also the relative importance of Agriculture and Fisheries as a direct and indirect supplier per $1 million of output of selected sectors in

[64] Food and Kindred Products intrasector activity as shown in Exhibit 14, column 2, row 2.

[65] In Exhibit 15, all amounts are considered significant only to the 5th decimal.

[66] Capital, labor, and inventory requirements for the 1947 agribusiness input-output data will be found in Appendix III, Section C.

TABLE 18. MAJOR INDUSTRIES THAT PROCURE DIRECTLY FROM AGRICULTURE AND FISHERIES, AND MAJOR INDUSTRIES THAT PROCURE DIRECTLY FROM THE BUYERS OF AGRICULTURE AND FISHERIES: 1947

Buyers from Agriculture and Fisheries	*Amount of Purchases*† *(Thousands)*	*Markets of the Buyers of Agriculture and Fisheries*	*Amount of Purchases*† *(Thousands)*
			FROM FOOD AND KINDRED PRODUCTS
FOOD AND KINDRED PRODUCTS	$15,048,000	Eating and Drinking Places	$3,469,000
		Agriculture and Fisheries	2,378,000
		Chemicals	685,000
		Leather and Leather Products	444,000
		Medical, Educ., and Nonprofit Orgs.	251,000
			FROM TEXTILE PRODUCTS
TEXTILE MILL PRODUCTS	2,079,000	Apparel	$3,882,000
		Rubber Products	444,000
		Furniture and Fixtures	285,000
		Motor Vehicles	147,000
			FROM CHEMICALS
CHEMICALS	1,211,000	Food and Kindred Products	$1,451,000
		Agriculture and Fisheries	830,000
		Textile Mill Products	800,000
		New Construction and Maintenance	635,000
		Rubber Products	604,000
		Medical, Educ., and Nonprofit Orgs.	222,000
		Products of Petroleum and Coal	213,000
		Personal and Repair Services	198,000
		Paper and Allied Products	183,000
		Other Electrical Machinery	178,000
		Miscellaneous Manufacturing	167,000
		Apparel	142,000
		Leather and Leather Products	126,000
		Stone, Clay, and Glass Products	116,000
		Motor Vehicles	111,000
			FROM EATING AND DRINKING PLACES
EATING AND DRINKING PLACES	865,000	Medical, Educ., and Nonprofit Orgs.	$152,000
TOBACCO MANUFACTURES	783,000		
			FROM LUMBER AND WOOD PRODUCTS
LUMBER AND WOOD PRODUCTS	192,000	New Construction and Maintenance	$2,330,000
		Furniture and Fixtures	385,000
		Paper and Allied Products	267,000
		Agriculture and Fisheries	148,000
		Rentals	135,000
MEDICAL, EDUC., AND NONPROFIT ORGS.	116,000		

† The values shown here are taken from Exhibit 13 expressed in "thousands of dollars" instead of "millions of dollars."

SOURCE: W. Duane Evans and Marvin Hoffenberg, "The Interindustry Relations Study for 1947," *The Review of Economics and Statistics,* May 1952.

TABLE 19. DIRECT AND INDIRECT REQUIREMENTS FROM SELECTED INDUSTRIES PER MILLION DOLLARS OF FINAL DEMAND OF THE MAJOR DIRECT AND INDIRECT BUYERS FROM AGRICULTURE AND FISHERIES: 1947

Direct and Indirect Buyers from Agriculture and Fisheries	*Direct and Indirect Dollar Inputs Required from Industries Listed Below for Each Million Dollars of Final Demand of Industries Named in the Opposite Column*† *(Thousands)*	
FOOD AND KINDRED PRODUCTS	Agriculture and Fisheries	$675
	Chemicals	90
	Trade	49
	Rentals	48
	Paper and Allied Products	41
	Railroad Transportation	38
	Products of Petroleum and Coal	33
	Transportation Other than Rail or Ocean	32
TOBACCO MANUFACTURES	Agriculture and Fisheries	$620
	Paper and Allied Products	76
	Business Services	63
	Trade	57
	Food and Kindred Products	57
	Chemicals	51
	Rentals	44
	Printing and Publishing	37
	Railroad Transportation	32
TEXTILE MILL PRODUCTS	Agriculture and Fisheries	$379
	Chemicals	141
	Trade	52
	Food and Kindred Products	44
	Paper and Allied Products	33
	Rentals	31
EATING AND DRINKING PLACES	Food and Kindred Products	$324
	Agriculture and Fisheries	274
	Trade	104
	Rentals	55
	Chemicals	39
	Railroad Transportation	38
	Coal, Gas, and Electric Power	37
	Personal and Repair Service	36
CHEMICALS	Agriculture and Fisheries	$209
	Food and Kindred Products	92
	Paper and Allied Products	76
	Products of Petroleum and Coal	70
LEATHER AND LEATHER PRODUCTS	Food and Kindred Products	$209
	Agriculture and Fisheries	169
	Chemicals	100
APPAREL	Textile Mill Products	$411
	Agriculture and Fisheries	148
	Chemicals	82
	Trade	63
	Miscellaneous Manufacturing Industries	31
	Paper and Allied Products	31

† The values shown here are taken directly from Exhibit 15, adjusted to a $1 million volume.

SOURCE: W. Duane Evans and Marvin Hoffenberg, "The Interindustry Relations Study for 1947," *The Review of Economics and Statistics,* May 1952.

TABLE 20. GROSS OUTPUT OF EACH PROCESSING SECTOR REQUIRED IN THE UNITED STATES ECONOMY TO SATISFY THE DIRECT PURCHASES OF FINAL CONSUMERS FROM EACH SECTOR: 1947

Sector	*Millions of Dollars*	*Direct (% of Total)*	*Indirect (% of Total)*
Agriculture and Fisheries	$37,208	26.3%	73.7%
Food and Kindred Products	33,352	66.4	33.6
Tobacco Manufactures	2,206	67.3	32.7
Textile Mill Products	7,389	19.9	80.1
Apparel	12,308	81.1	18.9
Lumber and Wood Products	1,769	3.8	96.2
Furniture and Fixtures	1,786	81.7	18.3
Paper and Allied Products	5,372	6.4	93.6
Printing and Publishing	4,933	30.2	69.8
Chemicals	9,104	21.6	78.4
Products of Petroleum and Coal	8,723	27.9	72.1
Rubber Products	1,778	39.9	60.1
Leather and Leather Products	3,278	63.0	37.0
Stone, Clay, and Glass Products	1,713	19.9	80.1
Iron and Steel	4,023	0.0	100.0
Nonferrous Metals	2,472	0.8	99.2
Plumbing and Heating Supplies	522	76.0	24.0
Fabricated Structural Metal Products	250	5.4	94.6
Other Fabricated Metal Products	3,200	16.8	83.2
Agricultural, Mining, and Construction Machinery	429	15.4	84.6
Metalworking Machinery	418	7.4	92.6
Other Machinery (except Electrical)	3,107	34.8	65.2
Motors and Generators	341	0.0	100.0
Radios	896	71.3	28.7
Other Electrical Machinery	1,924	35.0	65.0
Motor Vehicles	6,833	45.8	54.2
Other Transportation Equipment	681	25.1	74.9
Professional and Scientific Equipment	1,160	54.3	45.7
Miscellaneous Manufacturing Industries	3,323	58.2	41.8
Coal, Gas, and Electric Power	6,530	2.0	98.0
Railroad Transportation	5,972	34.5	65.5
Ocean Transportation	502	20.2	79.8
Other Transportation	7,153	54.0	46.0
Trade	33,545	80.8	19.2
Communications	2,492	50.9	49.1
Finance and Insurance Agents	11,127	62.8	37.2
Rentals	26,413	76.8	23.2
Business Services	3,753	4.8	95.2
Personal and Repair Services	11,630	63.0	37.0
Medical, Educational, and Nonprofit Organizations	8,148	96.4	3.6
Amusements	2,788	86.2	13.8
Scrap and Miscellaneous Industries	1,186	0.0	100.0
Nondistributed	15,962	0.0	100.0
Eating and Drinking Places	12,843	94.0	6.0

SOURCE: W. Duane Evans and Marvin Hoffenberg, "The Interindustry Relations Study for 1947," *The Review of Economics and Statistics*, May 1952.

relation to their requirements from *other* sectors in producing $1 million of final demand. One should note that even though Agriculture and Fisheries may be important to an industry sector per dollar of its final demand, the *total* final demand of such sector will determine its over-all importance as a market to Agriculture and Fisheries.

Still another type of analysis is illustrated by Table 20 which presents the relative importance of direct vs. indirect[67] input requirements created by final consumer demand (Households). The term direct requirements here refers to the values shown in column 50 of Exhibit 13. In reality they are the end-product items which move directly to the Consumers (Households) sector without intermediary processing. The indirect requirements, derived from Exhibit 15, include the additional output essential to support the activities generated by Consumers (Households). Because the values in Exhibit 15 are in terms of gross output, intrasector flows are included as a part of indirect requirements.[68]

It is interesting to note that of the inputs required to satisfy household demand, roughly half were direct and half indirect in terms of the overall picture for 1947. This relationship varies widely, however, as between specific sectors of the economy, as is highlighted by a comparison of the Agriculture and Fisheries sector, where each dollar spent by households generates approximately $3.00 of indirect transactions, with the Apparel sector, where each dollar of direct requirements is matched with about 20 cents of indirect requirements.

A word of caution is needed as to the effect of the method of aggregating upon analyses such as those presented in Exhibits 13, 14, and 15. For example, in the 1947 study all outputs of Coal, Gas, and Electric Power used by Households were charged to the Rentals sector. This means that these items appear as indirect requirements of households. Had they been dealt with in the reverse, then the ratio of direct to indirect requirements for the Coal, Gas, and Electric Power sector would have been altered materially.

The Bakery Products Industry

Thus far in this chapter we have examined the use of input-output economics in determining the general requirements of agribusiness and other sectors under stated conditions. We shall now examine two selected industries (Bakery Products and Meat Packing and Wholesale Poultry) in somewhat more detail.

By assuming a $1 million increase in the consumer purchases of bakery goods, and then tracing through the effect of this increase on both the various related sectors and aggregates of agribusiness and on the rest of the economy, one is able to note significant interrelationships relating to the agribusiness structure and visualize certain additional uses of the input-output technique.

Table 21, "Requirements for Producing $1,000,000 of Bakery Products,"[69] and Exhibit 16, derived from this table, illustrate in some detail the direct and indirect production required of each sector of the economy to support a $1 million output by the Bakery Goods Industry. Table 21 and Exhibit 16 indicate that $2,319,000 of transactions would be needed to produce such a quantity of

[67] Direct requirements as used in Table 20 refer to direct purchases by the final demand sector—consumer purchases (column 50, Exhibit 13)—from each of the processing sectors. Indirect requirements as used in Table 20 indicate processing sector activities required by the sector that supplies the direct requirements of consumers.

[68] The data in Table 20 were obtained directly from Exhibits 13 and 15. For example, the Agriculture and Fisheries sector had to supply $37,208 million of goods and services to the whole economy in order to satisfy the total direct purchases by Consumers (Households) from each of the processing sectors of the economy. To obtain the $37,208 million, one has to multiply each of the direct purchases by Consumers (Households) from each of the processing sectors (rows 1 through 44 of column 50, Exhibit 13) by each of the direct and indirect dollar production requirements from Agriculture and Fisheries (each of the same 44 sectors, rows 1 through 44, column 1, Exhibit 15). The percentage of direct purchases by consumers from Agriculture and Fisheries is obtained by dividing $9,785 million by $37,208 million (the direct purchases from Agriculture and Fisheries by consumers—see column 50, row 1, Exhibit 13) which equals 26.3%. The remaining 73.7% represents the indirect requirements from agriculture needed to support all the direct requirements of Consumers (Households) for all the processing sectors of the economy in 1947.

[69] The data for the Bakery Products sector were obtained from the 192 matrix table of indirect and direct requirements in order to illustrate in more detail its relation to the total economy. It is the same type of data that are presented in Exhibits 14 and 15 with the exception that in Exhibits 14 and 15 the Bakery Products sector is part of the combined aggregate "Food and Kindred Products" and that the amount required is per dollar of final demand instead of $1 million of final demand. This should not be confused with column 16 of Exhibit 2 (Bakery Products) which represents the direct purchases from the economy by Bakery Products in the year 1947 for its output of $3,353 million.

Table 21. Requirements for Producing $1,000,000 of Bakery Products: 1947
(Thousands of dollars)

	Direct	*Indirect*	*Total*
Food Processing Aggregate			
Grain Mill Products	$209	$29	$238
Miscellaneous Food Products	110	25	135
Sugar	33	65	98
Meat Packing and Wholesale Poultry	33	7	40
Vegetable Oils	6	29	35
Processed Dairy Products	28	5	33
Canning, Preserving, and Freezing	14	2	16
Alcoholic Beverages	—	3	3
Total	$433	$165	$598
Farming Aggregate			
Food Grains and Feed Crops	—	154	154
Meat Animals and Products	—	34	34
All Other Agriculture	—	26	26
Farm Dairy Products	6	18	24
Oil-Bearing Crops	—	20	20
Poultry and Eggs	5	6	11
Vegetables and Fruits	4	6	10
Total	$15	$264	$279
Container Aggregate			
Paper Products	39	14	53
Paper Mills	—	26	26
House Furnishings and Other Nonapparel	—	11	11
Spinning, Weaving, and Dyeing	—	8	8
Plastic Materials	4	2	6
Tin Cans and Other Tins	—	5	5
Glass	—	3	3
Logging	—	3	3
Total	$43	$72	$115
Transportation Aggregate			
Railroads	20	26	46
Automobile and Other Repair Services	14	6	20
Trucking	7	12	19
Motor Vehicles	—	11	11
Tires, Tubes, and Other Rubber Products	—	7	7
Water and Other Transportation	—	3	3
Total	$41	$65	$106
General Aggregate			
Real Estate and Rentals	4	32	36
Banking and Finance	3	10	13
Steelworks and Rolling Mills	—	7	7
Total	$7	$49	$56
Fuel and Power Aggregate			
Electricity	4	8	12
Petroleum Products	5	6	11
Coal Mining	—	8	8
Natural Gas	3	4	7
Total	$12	$26	$38
Fertilizer and Chemicals			
Chemicals	—	17	17
Fertilizer	—	5	5
Total	—	$22	$22
All Other	25	74	99
Bakery Products	1,005	1	1,006
GRAND TOTAL	$1,581	$738	$2,319

Source: Derived from data published by the Bureau of Labor Statistics, Division of Interindustry Economics.

EXHIBIT 16. DIRECT AND INDIRECT REQUIREMENTS FOR PRODUCING $1,000,000 OF BAKERY PRODUCTS: 1947

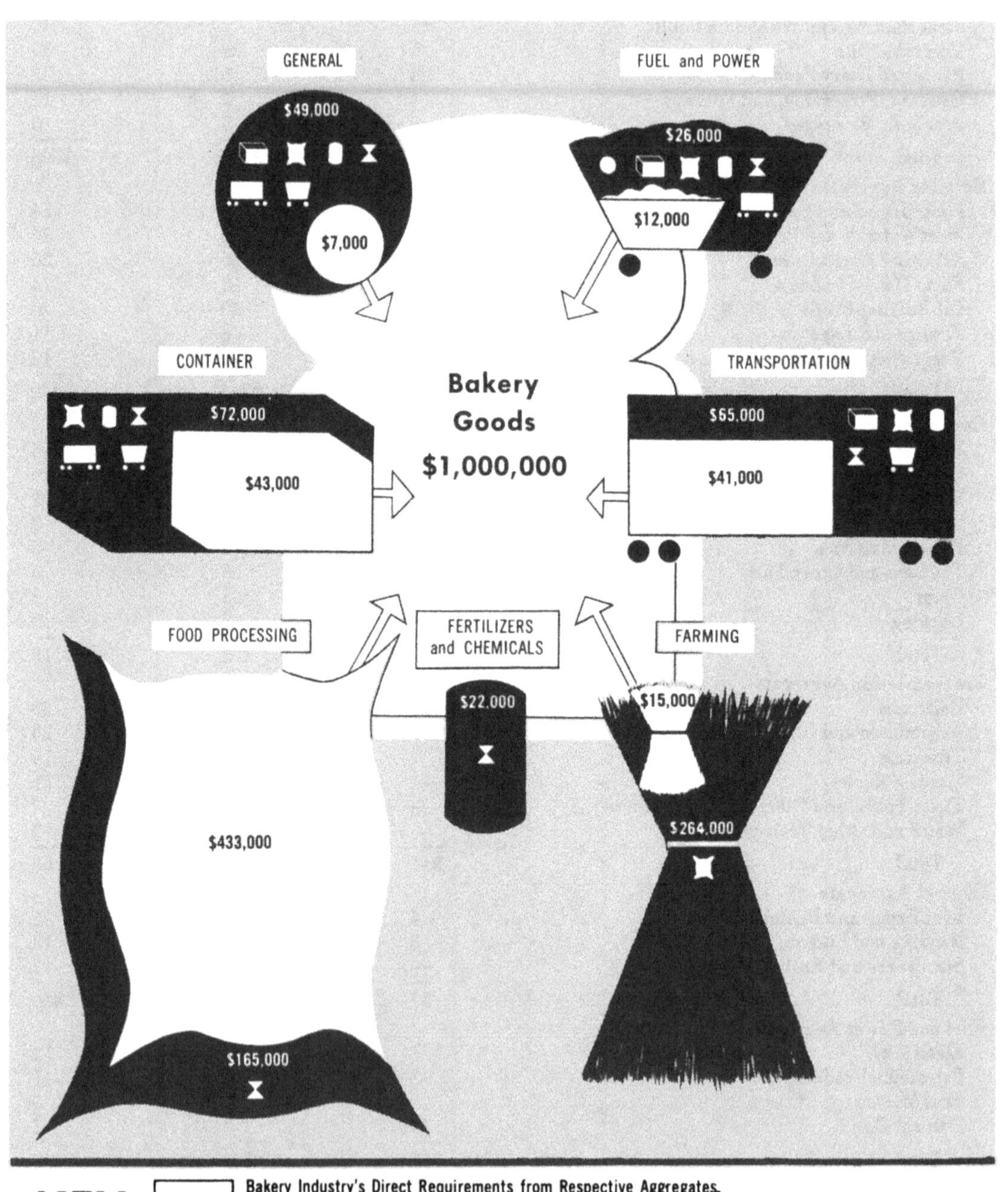

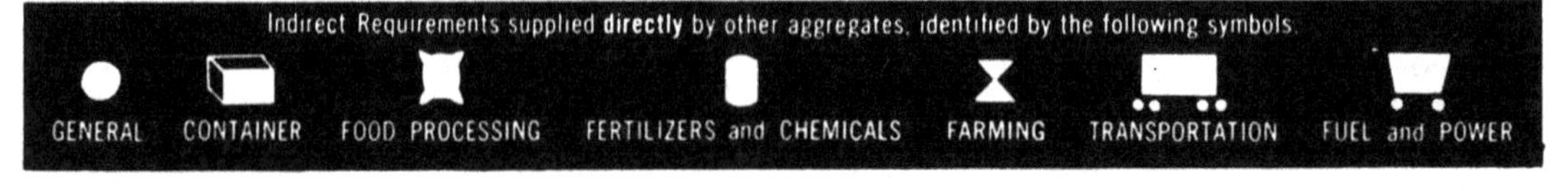

bakery goods. In other words, $1.00 of consumer demand would generate $1.32 of productive activity in other parts of the economy for every dollar spent directly on bakery goods.[70] Table 21 and Exhibit 16 are summarized below. Over four-fifths of the productive activity induced by the Bakery sector occurs within agribusiness—the direct purchases consisting of processed items such as flour, sugar, and dairy products, and the indirect, of products such as wheat, eggs, milk, fruits, and vegetables.

In other words, what happens over a bakery counter affects the direct and indirect suppliers of the bakery goods industry, such as the North Dakota wheat farmer or the fertilizer manufacturer, as well as the baker and his immediate customer. This is illustrative of agribusiness interrelationships.[71]

Requirements for Producing $1,000,000 of Bakery Products (in thousands)

Aggregate or Sector	*Direct*	*Indirect*	*Total*	*% of Total Requirements*
Food Processing Aggregate	$ 433	$165	$ 598	26%
Farming Aggregate	15	264	279	12
Fertilizer Sector	0	5	5	0
Bakery Products Sector	1,005	1	1,006	43
Main Agribusiness Sources	$1,453	$435	$1,888	81%
Container Aggregate	$ 43	$ 72	$ 115	5%
Transportation Aggregate	41	65	106	5
General Aggregate	7	49	56	2
Fuel and Power Aggregate	12	26	38	2
Chemicals Industry Sector	0	17	17	1
All Other	25	74	99	4
Subtotal	$ 128	$303	$ 431	19%
Total	$1,581	$738	$2,319	100%

The Meat Packing and Wholesale Poultry Industry

Table 22 shows the same type of data for the Meat Packing and Wholesale Poultry Industry as has been used to illustrate the requirements of the Bakery Products Industry. By comparing Tables 21 and 22, one is able to ascertain similarities and differences in the products required to support these two industries. The following abbreviated tabulation is derived from Tables 21 and 22 in order to facilitate direct comparisons:

Comparison of Requirements by Meat Packing and Wholesale Poultry Industry and Bakery Products Industry from the Farming Aggregate in Producing $1,000,000 of End Products (in thousands)

Purchases by Bakery Products Industry

Farming Aggregate Sales	*Direct*	*Indirect*	*Total*
Food Grains and Feed Crops	$ 0	$154	$154
Meat Animals and Products	0	34	34
Poultry and Eggs	5	6	11
Farm Dairy Products	6	18	24
Vegetables and Fruits	4	6	10
Oil-Bearing Crops	0	20	20
All Other Agriculture	0	26	26
Total	$ 15	$264	$279

Purchases by Meat Packing and Wholesale Poultry Industry

Farming Aggregate Sales	*Direct*	*Indirect*	*Total*
Food Grains and Feed Crops	$ 0	$403	$403
Meat Animals and Products	729	127	856
Poultry and Eggs	27	4	31
Farm Dairy Products	0	13	13
Vegetables and Fruits	0	8	8
Oil-Bearing Crops	0	3	3
All Other Agriculture	0	20	20
Total	$756	$578	$1,334

[70] The relationship of $1.00 of consumer demand to $1.32 of processing activity is obtained by comparing the $1 million of Bakery Products demand to the total requirements of $2,319,000 needed to produce the $1 million of demand (see Table 21). The amount of inputs required is in part a function of the detail and size of the matrix.

[71] The Capital, Labor, and Inventory positions of the Bakery Products sector are examined in Appendix III, Section D.

TABLE 22. REQUIREMENTS FOR PRODUCING $1,000,000 OF MEAT GOODS: 1947
(Thousands of dollars)

	Direct	*Indirect*	*Total*
Food Processing Aggregate			
Meat Packing and Wholesale Poultry	$1,050	$6	$1,056
Canning, Preserving, and Freezing	—	3	3
Grain Mill Products	—	46	46
Miscellaneous Food Products	5	4	9
Sugar	—	4	4
Vegetable Oils	—	6	6
Animal Oils	5	2	7
Total	$1,060	$71	$1,131
Farming Aggregate			
Meat and Animal Products	729	127	856
Poultry and Eggs	27	4	31
Farm Dairy Products	—	13	13
Food Grains and Feed Crops	—	403	403
Oil-Bearing Crops	—	3	3
Vegetables and Fruits	—	8	8
All Other Agriculture	—	20	20
Total	$756	$578	$1,334
Fiber Processing Aggregate			
Spinning, Weaving, and Dyeing	—	3	3
House Furnishings and Other Nonapparel	—	3	3
Total	—	$6	$6
Fertilizer Sector	—	$11	$11
AGRIBUSINESS TOTAL	($1,816)	($666)	($2,482)
Fuel and Power Aggregate	—	53	53
Transportation Aggregate	13	51	64
Container Aggregate	8	11	19
All Other	11	212	223
Subtotal	$32	$327	$359
GRAND TOTAL	$1,848	$993	$2,841

SOURCE: Derived from data published by the Bureau of Labor Statistics, Division of Interindustry Economics.

From the preceding tabulation one can see the validity of the view commonly expressed, that a meat economy would absorb more crop output than a cereal economy. Another fact worthy ot note is that the Bakery Products Industry requires a higher degree of processing of farm products than does the Meat Packing and Wholesale Poultry Industry sector, as indicated by the relationship of the indirect purchases of Bakery Products from Farming to its direct purchases ($264,000 to $15,000) and the relationship of the indirect purchases of Meat Packing from Farming to its direct purchases ($578,000 to $756,000).

CONCLUSION

In this chapter the authors have attempted to present in terms of 1947 data the many interrelationships that exist within the framework of agribusiness and to illustrate the significance of agribusiness to the remaining sectors of the national economy. This chapter also demonstrates the feasibility and utility of the input-output technique as a tool for analyzing our food and fiber economy. It makes possible the quantification of the reciprocal and rebounding movement of resources, goods, and services throughout the component phases of agribusiness and in relationship to the economy. In the future it should prove invaluable in finding new and more adequate answers to the mounting problems that trouble our food and fiber economy. While the merits of the input-output technique are not universally accepted[72] by economists and businessmen, the authors look upon it as a useful methodology for examining the complex relationships and interrelationships within agribusiness and between agribusiness and the rest of the economy.

Having examined in this chapter the inner workings of agribusiness for the specific year 1947, let us now turn to an appraisal of the general and specific possible uses of the agribusiness concept in the future.

[72] For a complete appraisal of the input-output technique see *Input-Output Analysis: An Appraisal*, Studies in Income and Wealth, Volume XVIII, by the Conference on Research in Income and Wealth, National Bureau of Economic Research, 1955.

CHAPTER 4

Agribusiness and the Future

THE PREVIOUS ANALYSIS has pointed out how the food and fiber portion of our economy has progressed from an agricultural to an agribusiness status during the past century and a half. Also, it has attempted to describe agribusiness as it exists today in terms of dimensions, magnitude, and composition and the flow of resources, goods, and services within agribusiness and between it and the national economy. In this chapter we shall consider an agribusiness approach to the solution of problems confronting the food and fiber part of the economy.

NEED FOR AN AGRIBUSINESS POLICY

In the opinion of the authors, the factor most limiting balanced progress and economic growth in the food and fiber structure of our economy today is the absence of a comprehensive, well-defined, and well-balanced agribusiness policy. In general, major interests, including farm leaders, businessmen, public officials, and even professional experts, have been slow to comprehend the magnitude of the evolutionary forces which are converting agriculture into agribusiness and *farm problems* into *agribusiness problems*. They have been slow to recognize the inevitable changes which are inherent in the application of improved technology—changes in size of units, organization, managerial competence, technical skills, capital requirements, market potentials, etc. They have been prone to view maladjusments as temporary deviations from a norm and have attempted to deal with them by means of short-run expediency measures rather than by recognizing them as basic changes in the very structure of our food and fiber economy—changes which will make the future unlike the past and which, accordingly, necessitate a constant re-evaluation of our food and fiber policy, taking into account all related functions, whether performed on or off the farm.

Of course, the need for re-examining our economic system with new perspective, oriented to changing conditions, is not limited to agriculture but is universal—both on a national and international scale. The point is that nationally we have done a better job of making this transition in most other major phases of the economy than in food and fiber. In general, nonagricultural business enterprises have tended to make decisions on the basis of the total operations of an industry—from preproduction determinations to the final sale of the products—to a greater degree than have the components of agribusiness. This is evidenced by the increasing size of business firms, the continuing trend toward vertical integration, the growth of trade associations, and similar coordinating efforts. While such developments also have charactized certain aggregates within agribusiness, it has not typified relationships generally as between aggregates.

Within the Department of Agriculture, the farm organizations and Congress policy deliberations have continued to be more or less farm-oriented, pertaining mainly to short-run considerations. By and large, little has emerged which points toward an adequate or permanent solution to the so-called farm surplus problem. Particularly notable has been the limited effort to build new strength in terms of economic stability into the private sectors of the economy as a backfire to the expanding role of government on the farm front. While such focusing of attention upon short-run phases of the problem is understandable and has a place, still no adequate solution to our food and fiber problems is likely to be forthcoming until such considerations are viewed in an agribusiness setting with long-range objectivity and against the background of our national economic objectives.

Even in the administration of the Full Employment Act of 1946, which created the Council of Economic Advisers to the President and the Joint Congressional Committee on the Economic Report, little has been done to assess properly the changing nature of our food and fiber economy or to relate it to the national economy. Such efforts have treated the production of food and fiber in the raw state as something more or less distinct in itself and only loosely related to business decisions affecting the availability and application of supply inputs or the postharvest decisions pertaining to the processing, fabricating, and distribution of the end products made from raw farm commodities. While such analyses have noted general trends in farm output, income, investment, employment levels, etc., they have not attempted to evaluate comprehensively the basic economic changes taking place in agribusiness or to interpret such effects upon the national economy.

Developing an Agribusiness Policy

The method of examining food and fiber problems in their comprehensive setting and developing adequate and objective policy consistent with national economic goals hereafter will be referred to as the agribusiness approach to policy formulation, and the resulting product will be labeled agribusiness policy.

With respect to the role of government in connection with economic stability, we would seem to have open to us three choices: (1) to continue to rely on governmental superstructure as a means of providing economic balance for agribusiness; (2) so to strengthen the economic sinews of agribusiness that it can provide economic balance for itself; or (3) to chart a middle course consisting of a combination of these two alternatives. In the discussion that follows, the authors assume the need for economic stability geared to progress in agribusiness and the presence of a desire to shift responsibility for such from government to the private sectors of the economy as rapidly as possible and to the degree possible. Of course, implicit in this objective is the further assumption that the rate and degree of transfer of responsibility will be compatible with the optimum achievement of our national economic goals. While past experience strongly argues that heavy reliance on government is not conducive to optimum balance and progress, particularly when it involves the ownership of surplus stocks, still with government already as much involved in agribusiness as it is today, time will be needed in which to effect the building of new economic strength in agribusiness and thus to lay the groundwork for the gradual transfer of responsibility to private enterprise. Hence, for the immediate future at least, we have little choice but to proceed in terms of alternative three, even as a means of ultimately achieving alternative two.

In determining the role of government we must not be guided by a blind zealousness to get government entirely out of agribusiness at any cost, but by the principle of putting responsibility where it best can be exercised in the national interest. Candidly appraised, it seems likely that the role of government will continue to be relatively large even for the long pull. The important thing is that we do not decide by default for it to be larger than it need be, because we failed to see problems in their true perspective or because the private sectors of the economy failed to act on opportunities open to them—opportunities which would have been remunerative and added strength to the food and fiber economy.

The agribusiness approach can be applied either to broad policy issues or to lesser considerations. In any case the essential thing is to examine the given issue in its comprehensive setting. In a dynamic economy, policy formulation must be a continuous process—such being the basis of progress. Therefore, the agribusiness approach to a food and fiber policy must be evolutionary in nature, simultaneously taking into account both national considerations and the basic differences between commodities and regions. This means that problems must be examined both in their vertical setting with respect to the production,[1] processing, and distribution peculiarities of the given commodity, and in their horizontal setting with respect to the policies of other commodities, the national economy, and our international responsibilities. Thus, agribusiness policy-making consists of appraising problems first at the commodity and/or regional level and then coordinating the numerous efforts of this type into a general national policy.

The Role of Research

The starting point in formulating sound agribusiness policy is objective and comprehensive research for the purpose of developing facts and

[1] Including the manufacture and procurement of production supplies.

analyzing alternatives. In the past, public agencies and private firms have engaged extensively in research pertaining to specific problems such as killing insects, combatting diseases, developing better plants or animals, or ascertaining the cost of performing given farming or marketing functions. But there has been a void of adequate comprehensive studies designed to provide policy makers with essential facts and alternatives even for a single commodity, much less on a national scale. The result is that farm organizations, commodity interests, and business associations frequently adopt resolutions which are hastily and poorly thought out. Furthermore, little effort is made through the objective application of research to reconcile differences between groups or to work out the best policy, drawing on the ideas of all interested groups as source material.

In a climate of dissension and bickering such as exists from time to time on the rural front it is difficult for research organizations, particularly publicly financed agencies, to take the initiative in relating research to policy formulation. The key to success lies in developing confidence in research findings in the minds of agribusiness policy makers. Probably no better means exists for establishing such confidence than for a few well-planned, agribusiness-oriented research studies to be made by highly reputable institutions. Even if at first the findings of such studies tend to be ignored or disputed, in time they will gradually gain acceptance—provided, of course, the findings actually have validity.

An important step in the direction of better research is closer teamwork at the university level between colleges and schools of agriculture and business administration. In fact, the real need is for agribusiness experts, particularly economists and analysts, who have a breadth of competence sufficient to deal adequately with economic problems relating to food and fiber in their total setting—including both farm and off-farm phases. An important duty of such experts would be the training of additional research workers and teachers for coping with agribusiness problems. In turn, agribusiness orientation should be given to the agricultural extension service, vocational agriculture, and similar programs. This, of course, would not mean the elimination of technical education relating specifically to farming, but rather supplementing this training with appropriate information which would help the student also to see farming in its relationship to business functions. The timing element in research also is highly important both with respect to putting first things first and with respect to achieving public acceptance. Researchers constantly need to look ahead for the purpose of filling in the gaps which currently exist as the result of a piecemeal approach in the past, and to give high priority to such projects. The development of a broad, farsighted agribusiness research program will require courage to overcome the inherent hesitancy and timidity to tackle research in controversial areas, which has characterized our past performance. Also, researchers of the highest competence and integrity will be required for the development of such an agribusiness research program.

The Art of Policy Formulation

A second step in the formulation of agribusiness policy is that of developing a sound course of action on the basis of adequate facts. Fundamentally, this is a task for spokesmen for the respective segments of agribusiness, including farm organizations, commodity groups, business associations, public administrators, and legislators, assisted by researchers and technicians who can provide them with facts and help them to analyze such data objectively. This policy formulation, of course, will entail collaboration among such interests of a higher order than we have had heretofore. If conferences of this type are to produce results, two things are of prime importance with respect to the attitude of the conferees: (1) an open mindedness toward the truth, even when it disproves a basic tenet of an organization, and (2) a willingness on the part of organizations and individuals to subordinate a group interest to the broader interest of agribusiness and the nation when conflict develops. In the long run the pattern of each group fighting for a special advantage is a costly one in terms of progress and national welfare. All too often a controversial problem drags on unsolved until conditions reach emergency proportions. Then, in desperation, interested groups call upon the administrative arm of government or Congress to intervene. Frequently, under the pressure of impending crisis and in the name of expediency, an inadequate, compromised answer is improvised. Such an answer is all too likely to be second- or third-rate, unnecessarily saddling on government a responsibility which private interests better might have assumed had the various groups worked together in a sincere and honest effort to achieve a sound answer.

The art of getting diverse and even contentious groups to work together in the development of an agribusiness policy will not be easy. However, a beginning already exists as the result of numerous conferences which have been sponsored by one or more of the participants, and others by educational institutions, civic organizations, and foundations. In addition, during recent years, the United States Department of Agriculture has held meetings of advisory committees, representing virtually every commodity, composed of members from agriculture and business. In general, these sessions have been for the purpose of exchanging ideas on specific problems rather than establishing a general policy. However, the significant fact is that the leaders of agriculture and business have had experience working together. The task now is to get them to cooperate in formulating an agribusiness policy, using the agribusiness approach.

Formulators of agribusiness policy will need patience, particularly during the initial efforts. When agreement cannot be reached at a given time, it may be necessary to table the issue until further research can be done. Also it is important that policy be developed in a systematic manner through a democratic procedure which is conducive to full discussion and mutual education. As already indicated, the formulation of agribusiness policy must be an evolutionary process that builds on established areas of agreements by refining, improving, and extending them. Insofar as possible, solutions should be sought which are beneficial to the long-run interests of all agribusiness groups and the nation as a whole. Encouraging in this respect is the fact that, generally speaking, the longer run interests of farmers and businessmen do tend to run parallel.

At least in the beginning stages, collaboration between groups probably can best take place informally, without any attempt to create a new super organization. Coordinating committees, task groups, and *ad hoc* working parties can play an important role, provided due care is exercised in selecting leadership and provided the participating members actually desire to achieve results.

Even within government there is need of interdepartmental coordination on an agribusiness basis. This is true because matters concerning food and fiber not only involve the Department of Agriculture, but also frequently concern the departments of Commerce, Interior, Treasury, State, Defense, the Bureau of the Budget, and others. This need for teamwork has become increasingly apparent in recent years as farm problems have gained in prominence. Actually informal machinery already exists in government for a certain amount of coordination. This can serve as a starting point for developing more adequate procedures, measured in terms of the timeliness and quality of decisions.

Integrating Research and Policy-Making

In the background of agribusiness, as in all phases of our economy, there is need for basic research. Much of this, at the time, may seem only remotely related to current problems and issues. This type of research is fundamental if sound, continuing progress is to be made. Technicians working in such areas need to be given wide latitude and freedom from the economic and political pressures of the day in order that they may be fully objective in their approach and findings. In the opinion of the authors, this type of research needs to be developed more fully with respect to the economic, social, and organizational phases of agribusiness.

Concurrently there also is need for types of semibasic and applied research which are more directly related to policy formulation. These types of research are essential in bridging the gap between basic principles and courses of action. It is in this phase particularly that we have been weak in the past. The segmented nature of agribusiness and the multiplicity of operating units within the food and fiber phase of the economy have been major contributing factors to this weakness.

If a sound agribusiness policy is to be developed in the future, then those responsible for such a policy must approach problems with greater objectivity. This means that they must have a high appreciation of the role of research. Such leaders must seek true facts and honest answers, even when this means departing from the policies of the past.

Those engaged in research, both basic and applied, must be persons possessing competence, originality, and integrity. While with respect to the applied phase researchers need to be working closely with policy makers in order that informational gaps may be filled, still they must be sufficiently independent to present truths as they find them. This need for objectivity means that much of the applied research, particularly that dealing with controversial issues, must be done outside the farm organizations, business firms, and governmental agencies which participate in policy formulation. A useful technique in this respect is that of contract research whereby policy makers seeking factual information on a given problem engage an independent research entity to undertake the study.

In many phases of agribusiness it probably will be necessary to start with a survey or reconnaissance study designed to explore and outline the type of follow-up research that is needed. For example, if one were to undertake a project oriented by an agribusiness approach in the area of livestock marketing, numerous preliminary questions would need to be answered, such as: Should it cover all types of meat (including beef, pork, lamb, and mutton), poultry, and fish? Should it include all phases of operation from production to retailing? Should it consider competitive or substitute foods? What should be the emphasis with respect to the various factors considered in the study? Instead of one comprehensive project, should the research be broken down into a series of related efforts? If so, what types of research institutions should undertake each phase? What methodology is most appropriate for each phase of research? What place has the input-output technique as a device for portraying and quantifying relationships? These and numerous other questions need to be answered before new research efforts are undertaken on an agribusiness scale.[2]

[2] The authors have in mind studies of this exploratory type dealing with livestock and/or cotton as a follow-up to this general study.

Antitrust Laws

In discussions relating to the supply or price of farm commodities, the agribusiness approach may run into conflict with present antitrust laws and regulations. Agribusiness policy makers should expect to adhere to the spirit of the antitrust provisions. The development of agribusiness policy, however, is likely to raise new issues which will need clarification. For example, is it contrary to the spirit of antitrust laws for farm and business leaders to collaborate in planning a course of action which will shift to private enterprise some or all of the responsibility for economic stability in agriculture which government now assumes? If the answer is yes, then is it likely that we will ever reduce the role of government in agriculture? If the answer is no, then how far can leaders of agriculture and business go in their collaboration?

Some guidance in answering these questions can be found in the Capper-Volstead Act and in the antitrust provision of the Agricultural Adjustment Act pertaining to marketing agreements. As a general principle, the authors suggest modification of antitrust provisions to the degree necessary to place agribusiness on an economic basis comparable with that of business generally,[3] taking into account the inherently weak position peculiar to agricultural management with respect to relating supply and demand. More specifically, this modification of antitrust provisions would mean giving latitude to agribusiness components sufficient to enable them to overcome the disadvantages that have given rise to government price-support programs. In no sense, however, should it mean giving agribusiness a preferential position, in fact, relative to the economy in general. In this connection it is important that antitrust provisions be appraised comprehensively and with agribusiness perspective for the purpose of achieving equity rather than in the light of legalistic concepts which tend to have the effect of placing the food and fiber economy in an unrealistic legal straightjacket for the purpose of achieving superficial administrative consistency.

[3] This does not imply that present antitrust provisions relating to business represent an ideal—merely that agribusiness and other business interests should be on a common basis.

Content of an Agribusiness Policy

The opportunity for solving the so-called farm problems by the agribusiness approach to policy is made more promising by virtue of the fact that agribusiness evolution still is taking place in a national economy that is rapidly expanding. As was pointed out in the introduction, the so-called farm problems tend to gravitate about two poles—one pertaining to "commercial farming" and the other, "low-income farming." Let us now consider each of these in terms of formulating an agribusiness policy.

Commercial Farming

The answer to the problems of commercial farming is progress. Food and fiber still satisfy man's most basic needs. If these needs are to be supplied adequately with respect to quantity, quality, and value, then agribusiness must keep pace with the rest of the nation in terms of economic incentives and productivity. Equally important is balanced progress on the social, cultural, and spiritual phases of rural life.

Today the problems of commercial agriculture are growing in complexity and magnitude—almost to the point of appearing to defy solution. But they are not insoluble. As has been indicated repeatedly, such problems need to be approached as agribusiness issues because both their cause and their solution encompass the off-farm functions of supply manufacturing and processing-distribution as well as on-farm production. *The point is that the approach to solutions must be as comprehensive as is the bases of the problems themselves.*

Among the items which need to be reappraised are such questions as the short- and long-run needs of the United States for food and fiber; the future of the family-operated farm; the minimum size of a farm from the standpoint of efficiency; capital needs and capital sources; the responsibility of supply firms with respect to over-production; and the responsibility of processors and distributors for market expansion and price behavior. From the standpoint of future progress in the food and fiber part of the economy, probably no other issue confronting commercial agriculture is as vital as finding a satisfactory solution to the *cost-price squeeze.* Failure to solve this problem satisfactorily will lead to continuing pressure for price-support programs and other forms of governmental aid which tend to deal with short-run effects rather than causes and hence to retard true progress.

Because the cost-price squeeze is a complex problem brought about by a number of factors, on and off the farm (explained in Chapter 2), responsibility for finding a solution must be shared jointly by farmers, research agencies, the manufacturers of farm supplies, and processor-distributors, since all are helping to increase the capacity to produce food and fiber and/or to determine the size of the farm market. During the interim, while an agribusiness policy is being developed, we have little choice but to utilize government programs as a means of obtaining time in which to evolve better answers. Because of the great variation between commodities with respect to the nature of the problem, much of the basic work in quest of answers will involve using the agribusiness approach at the commodity level. As previously indicated, this approach entails analyzing the problem vertically in terms of production, processing, and distribution, and horizontally with respect to other commodities and the national economy.

The experiences of the past 30 years provide a valuable base for building agribusiness policy by providing examples of techniques which have worked and which have not worked. Particularly worthy of careful study are agribusiness devices involving producer-business teamwork such as the plentiful food campaigns, the work of the National Cotton Council, the American Dairy Association, and the Livestock Meat Board. Other programs involving more formal integration include marketing agreements and marketing orders. Of still another type is the vertical integration achieved through the cooperative ownership by farmers of off-farm facilities for handling supplies and for processing and distributing farm commodities. Far less common, but still illustrative of a type of integration, are the examples of the large farming unit which does its own processing and marketing and the food processing establishment which produces its own commodities. All such efforts involve agribusiness-type teamwork in the relating of on-farm and off-farm functions. While projects of this type have not proven adequate as solutions to current problems affecting commercial agriculture, they do represent significant joint efforts in relating on- and off-farm operations on which we can build for the future. In a sense they are "breakthroughs" into the agribusiness frontier. As such they are but beginnings to the solution of problems—not final answers. But they are experiences from which agribusinessmen can glean ideas. They are stepping stones toward an agribusiness policy.

There is strong tradition in America for retaining the family farm as the dominant production unit in commercial agriculture. While in recent decades the number of such farms has decreased and the size increased, still in 1954 approximately 98% of all farms were owned by individuals and 75.6% were owner operated. In formulating agribusiness policy, special consideration should be given to the effect of any given proposal on the family farm as a production unit. If such farms are to retain their status in competition with other types of land ownership and management, effort must be made to sustain an environment which will enable the family farm to adjust successfully to changing conditions in terms of size, capitalization, improved technology, and essential management skills.

Low-Income Farm Families

The answer to the problem of low-income farm families is also one of progress—progress in the direction of enabling each individual to realize his optimum opportunity in terms of productivity and earning power. As explained in Chapter 2, while some such individuals may find their best opportunity in farming, the majority must either seek to supplement their farming activities with supplemental off-farm employment or turn to full-time off-farm work.

Here, as in the case of commercial agriculture, there is need for an agribusiness policy. Much of the answer lies off the farm with business, educational institutions, civic organizations, and employment agencies in connection with training persons for more productive work and providing employment opportunities. The place of research in the formulation of agribusiness policy pertaining to low-income families is as basic as in the case of commercial farming. Here, too, the process of formulating policy involves starting with the present and gradually evolving improvements through time. Policies must vary by regions and age groups in order to be most effective. The prospect is that farm families will continue indefinitely to rear more children than can find employment in agriculture. Hence, it follows that the problem of assisting farm youth to train for and find off-farm opportunities will be a permanent one, even though gradually its magnitude should decrease as the lag period in implementing adjustments is reduced.

Some public assistance doubtless will be needed to help aged and handicapped persons, provide vocational training for those farm workers who desire to shift to other employment, and supply vocational guidance and aid in obtaining employment. Also, at least for the present, a form of rehabilitation credit may be essential for use in assisting families to make the transition to more productive employment opportunities.

The program to aid low-income families launched by the United States Department of Agriculture is a long overdue start in attacking this problem and needs to be followed through aggressively.

Agribusiness Policy and Progress

As had been indicated repeatedly, the agribusiness approach to the problems of food and fiber is not in itself another farm program or even a solution to a farm problem, but a means to a solution. The heart of it is the viewing of such problems in their combined on-farm and off-farm setting and the blending together of many partial answers into a national agribusiness policy—one which, in turn, harmonizes with and reinforces our national economic objectives. The agribusiness approach must be an evolutionary process which is geared to progress. It must proceed by improving the tens of millions of decisions made each day by several million farmers and several thousand businessmen on the basis of enlightened self-interest. Because these decisions can only be based on the knowledge possessed by the persons making them, it is essential that all policy makers consistently have available to them information that is adequate and up to date. This need for up-to-date information, in turn, necessitates an extensive and comprehensive research system that is able to penetrate the agribusiness frontier in a systematic and orderly manner for the purpose of finding new answers to pressing problems. But even such a research system will not provide an adequate solution to food and fiber problems unless the leaders of the respective phases of agribusiness also are willing to work together in enlightened self-interest for the purposes of resolving differences and coordinating efforts.

Nor is the agribusiness approach offered as a panacea or a road to Utopia. It is not. At best it will be imperfect in the same way that all human endeavors are imperfect. Also, because of its comprehensiveness and complexity, progress will come slowly—particularly during the initial stages. But the very fact that the task of formulating farm policy will tend to become more and more complicated as technology increases leaves us with little choice but to tackle the job aggressively now, if we want our food and fiber economy to be dynamic, progressive, productive, and appropriately remunerative to those engaged in it. For the long pull the agribusiness approach offers new hope of satisfactorily solving "farm problems" in a manner consistent with American traditions and economic philosophy. The all-important thing, of course, is really to look for the cause of each problem and then to concentrate on a sound solution. The time has come to cease resorting to stopgap emergency measures which have a way of becoming permanent policy because of the lack of better answers.

Appendices

APPENDIX I

Glossary

Advertising includes all purchased advertising services and maintenance of advertising firms.

Aggregate includes groups of individual sectors and industries.

Agribusiness is the sum total of all operations involved in the manufacture and distribution of farm supplies; production operations on the farms; and the storage, processing, and distribution of farm commodities and items made from them.

Agriculture Aggregate (see Farming Aggregate)

Alcoholic beverages includes malt liquors, malt, wines and brandy, and distilled liquors.

All other agriculture includes tree nuts, legumes and grass seeds, sugar and sirup crops, miscellaneous crops, forest products, greenhouse and nursery products, and agriculture services.

Animal oils includes marine animals' oils, grease, tallow, fatty acids, bone meal, and tankage. Lard is not included.

Apparel includes dressed and dyed furs and all apparel except custom tailors and dressmakers.

Automobile and other repair services includes auto repairs and services performed in establishments primarily engaged in furnishing auto repair, storage, washing, rental, and parking services to the general public; and services of auto, tire, battery, and accessory dealers, and filling stations.

Bakery products includes bread and other bakery products, biscuits, crackers, and pretzels.

Canning, preserving, and freezing includes canning sea foods, curing fish, canning and preserving food, dehydrating fruits and vegetables, pickling and making sauces, and freezing foods.

Canvas products includes all products made of canvas such as sails, bags, and tarpaulins.

Consumers (Household Column in Evans, Hoffenberg Study) includes sales of goods and services to individuals or final consumers.

Container Aggregate includes paper and board mills, metal container materials and cork products, glass, tin cans and other tin ware, and wood containers.

Cotton includes the growing and sale of raw cotton only.

Eating and drinking places includes all restaurants, cafeterias, and bars.

Electric light and power includes the generation, transmission, or distribution of electrical energy for sales; and the production and distribution of steam for sale for heating purposes.

Exports minus competitive imports includes foreign purchases of the United States minus all cost to the U.S. Government of imports, travel abroad, etc.

Farm Supplies Aggregate includes all industries that manufacture and distribute farm production supplies.

Farming Aggregate includes meat animals and products, poultry and eggs, farm dairy products, food grains and feed crops, cotton, tobacco, oil-bearing crops, vegetables and fruits, all other agriculture, and fishing, hunting, and trapping. The fishing, hunting, and trappings sectors of the Farming Aggregate are excluded as part of the Farming Aggregate in every table, graph, and discussion with the exception of the input-output matrices.

Farm dairy products includes all dairy products produced on the farm.

Fertilizers includes superphosphates, mixed fertilizers, and some sulfuric acid.

Fiber Processing Aggregate includes apparel; spinning, weaving, and dyeing; leather and leather goods; house furnishings; special textile products; jute, linens, cord, and twine; canvas products.

Fishing, hunting, and trapping includes commercial fisheries and commercial hunting and trapping.

Food grains and feed crops includes corn, oats, barley, rye, wheat, and rice.

Food Processing Aggregate includes meat packing and wholesale poultry, miscellaneous food products, grain mill products, processed dairy products, bakery products, canning, preserving, and freezing, alcoholic beverages, tobacco manufactures.

Glass includes flat glass, glass containers, pressed glass, and blown glass. Optical lenses are not included.

Government purchases includes sales of goods and services to federal, state, and local governments. The sales to government-operated and -owned plants are shown as sales to the appropriate industry classification.

Government services includes payments to federal, state, and local governments (taxes, social security, import duties, postage, and miscellaneous). The purchases of government-owned plants are included under the appropriate industry.

Grain mill products includes flour and meal, prepared animal feeds, cereal preparations, rice cleaning and polishing, and blended and prepared flour.

Gross domestic outlays includes total purchases of each sector or aggregate of the economy from the other sectors as defined for the Agribusiness Interindustry Chart.

Gross domestic output includes total sales of each sector or aggregate of the economy to the other sectors as defined for the Agribusiness and National Economy Chart.

Gross private capital formation includes outlays for a new plant and capitalized equipment.

Gum and wood chemicals includes products of hardwood distillation, softwood distillation, gum naval stores, and natural dyeing and tanning materials.

House furnishings and other nonapparel includes textile bags, textile products not elsewhere classified, hard surface floor coverings except wood, rubber, or asphalt floor tiles, coated fabrics (except rubberized), curtains and draperies, and house furnishings not elsewhere classified.

Imports (noncompetitive) includes goods imported into this country which do not compete with the products of the purchasing industry.

Inventory changes includes increases or depletions of finished products only.

Jute, linens, cord, and twine includes manufacture of flax yarn and thread, fabrics of linen and cotton mixtures, jute yarn, jute goods (except felt), cotton cordage, and twine.

Labor and capital consumption (Households row in Evans, Hoffenberg Study) includes wages, salaries, interest, donations, management services, and depreciation.

Leather and leather goods includes leather tanning, other leather products, and footwear (excluding rubbers).

Maintenance construction includes all repair and maintenance construction.

Meat animals and products includes meat animals and other livestock, and livestock products.

Meat packing and wholesale poultry includes meat packing, wholesale and custom slaughtering, prepared meat manufacture, poultry and small game dressing, and wholesale lard.

Medical, dental, and other professional services includes medical, surgical, and other health services, hospitals (state, local, federal, and veterans' administration), and other professional services (lawyers, accountants, engineers, architects, veterinarians, etc.).

Metal container materials and cork products includes metal shipping barrels, drums, and kegs.

Miscellaneous food products includes confectionery products, chocolate and cocoa products; chewing gum; bottled soft drinks; liquid, frozen, and dried eggs, leavening compounds; shortening and cooking oils; vinegar and cider; macaroni and spaghetti; manufactured ice; and food preparations.

Oil-bearing crops includes soybeans, peanuts, linseed, and cottonseed.

Other chemical industries includes inorganic chemicals, organic chemicals, and miscellaneous chemical industries.

Other fuels (coal, coke, and gas) includes cyclic coal-tar crudes, beehive and by-product coke, and production of anthracite and bituminous coal, natural gas, transmission and distribution, manufacture and distribution of illuminating gas, and the distribution of mixed gas.

Paints and allied products includes paint, varnishes, and inorganic color pigments.

Paper and board mills and converted paper products includes mills engaged in the manufacture of paper, paper board, building paper, and building board from wood pulp and other fibers on a standard or modified paper machine, and coating and glazing paper, and converting paper or board whether purchased or produced by the same plant.

Petroleum products includes petroleum refinery products, lubrication oils and greases not made in petroleum refineries, and products of petroleum and coal not elsewhere classified (major products in 1947 were calcined petroleum coke and compounded or blended wax).

Poultry and eggs includes producers of poultry and eggs.

Primary Triaggregate includes the three main dimensions of agribusiness—The Farm Supplies, Farming, and Processing-Distribution Aggregates.

Processed dairy products includes all milk and cream sold by processing plants to distributors; and the manufacture of butter, cheese, and other dairy products; and ice cream made by retailers.

Processing-Distribution Aggregate includes all industries that manufacture, store, assemble, and distribute farm commodities and items made from them.

Railroads includes line-haul operating railroads, switching and terminal companies, sleeping car and other passenger car service, railway express service, and rental of railway cars.

Real estate and rentals includes real estate agencies; farm and urban, residential and business rents.

Retail trade includes retail and service establishments.

Sales within the sector or aggregate consists of purchases of goods by one establishment from another establishment within the same sector or aggregate.

Saw mills, planing, and veneer mills includes sawmills, planing, veneer, shingle, cooperage stock, and excelsior mills; wood preserving; and logging operations reported in conjunction with sawmills.

Secondary Triaggregate includes the Farming, Food Processing, and Fiber Processing Aggregates.

Sector represents specific industry groups in agribusiness and national economy (Exhibit 2). It also represents aggregates of industries in the Evans & Hoffenberg Charts, see Exhibits 13, 14, and 15.

Soap and related products includes soap, synthetic detergents, cleaners, sulfonated oils, and crude and refined glycerin.

Special textile products includes wool carpets, rugs, felt goods, lace goods, paddings and upholstery fillings, processed textile waste, and textile goods not elsewhere classified.

Spinning, weaving, and dyeing includes yarn throwing mills, thread mills, cotton broad mills, rayon and related broad-woven fabric mills, narrow fabric mills, and dyeing and finishing textiles (except woolens and worsted textiles and knitted goods).

Sugar includes manufacture of raw sugar, sirup and/or finished cane sugar from sugar cane, cane sugar refining, and beet sugar production.

Tin cans and other tins includes packers' cans, beer cans, general tin cans, oil containers, milk and ice-cream cans, and other tin ware (except household and hospital utensils) from purchased tinplate, terneplate, blackplate, or enameled sheet metal.

Tires, tubes, and other rubber products includes pneumatic casings, inner tubes, and solid and cushion tires for all vehicles using tires greater than two inches, rubber footwear, reclaimed rubber, industrial and mechanical rubber goods, rubberized fabrics, vulcanized rubber clothing, and tire retreading.

Tobacco includes the growing of tobacco only; it does not include processing tobacco.

Trucking includes for-hire trucking service; warehousing and storage activities of trucking firms are not included.

Vegetable oils includes the extraction of cottonseed oil, linseed oil, soybean oil, and their by-products. It does not include establishments primarily engaged in refining edible oils. (Shortening and cooking oils are included under "miscellaneous food products.")

Vegetables and fruits includes fruits (citrus fruits, eight major deciduous fruits, and six minor fruits), berries, vegetables (commercial for fresh marketing and processing), and melons.

Warehouse and storage includes warehousing, storage, forwarding, arrangement of transportation, and stockyards.

Water and other transportation includes freight and passenger service on the sea by United States companies to foreign ports, transportation on the water other than overseas transportation, toll ferries, operation of docks and piers, and stevedoring services; domestic and foreign air transportation, airport and flying field operations, and maintenance; local transit, electric interurban transit, and highway transportation excluding trucking.

Wholesale trade includes all wholesale trade except manufacturers' sales offices, retail establishments' wholesale trade, and goods sold by agents and brokers whose fees are not included in the cost of goods.

Wood containers and cooperage includes the manufacture of fruit and vegetable baskets, rattan and willow ware (except furniture), cigar boxes, wooden boxes, and cooperage.

The INTERINDUSTRY RELATIONS STUDY is based on the 1947 CENSUS OF MANUFACTURES. Data for the Census were collected from industry on the basis of establishments, and firms were not allowed to aggregate reports for more than two establishments in any one county. Thus, captive consumption occurring between establishments within a company has been reported for the most part, while captive consumption within an individual establishment may not have been reported. Figures may not add to total of 100% because of rounding.

All Other Industries Purchases

Includes the Purchases of the Following Industries

Other nonmetallic minerals
Logging
Plywood
Fabricated wood products
Wood furniture
Partitions, screens, shades, etc.
Pulp mills
Paper and board mills
Converted paper products
Printing and publishing
Industrial inorganic chemicals
Industrial organic chemicals
Plastics materials
Synthetic rubber
Synthetic fiber
Explosives and fireworks
Drugs and medicines
Fertilizers
Miscellaneous chemical industries
Petroleum products
Coke and products
Paving and roofing materials
Glass
Cement
Structural clay products
Other miscellaneous nonmetallic minerals
Pottery and related products
Concrete and plaster products
Abrasive products
Asbestos products
Blast furnaces
Steel works and rolling mills
Iron foundries
Steel foundries
Primary copper
Copper rolling and drawing
Primary lead
Primary zinc
Primary metals, n.e.c.
Nonferrous metal rolling, n.e.c.

n.e.c. = not elsewhere classified

Primary aluminum
Aluminum rolling and drawing
Secondary nonferrous metals
Nonferrous foundries
Iron and steel forgings
Tin cans and other tin ware
Cutlery
Tools and general hardware
Hardware, n.e.c.
Metal plumbing and vitreous fixtures
Heating equipment
Metal furniture
Structural metal products
Boiler shop products and pipe bending
Metal stampings
Metal coating and engraving
Lighting fixtures
Fabricated wire products
Metal barrels, drums, etc.
Tubes and foils
Misc. fabricated metal products
Steel springs
Nuts, bolts and screw machine products
Steam engines and turbines
Internal combustion engines
Farm and industrial tractors
Farm equipment
Construction and mining machinery
Oil-field machinery and tools
Machine tools and metalworking machinery
Cutting tools, jigs and fixtures
Special industrial machinery
Pumps and compressors
Elevators and conveyors
Blowers and fans
Power transmission equipment
Industrial machinery, n.e.c.
Commercial machinery and equipment, n.e.c.
Refrigeration equipment
Valves and fittings
Ball and roller bearings
Machine shops
Wiring devices and graphite products
Electrical measuring instruments
Motors and generators
Transformers
Electrical control apparatus
Electrical welding apparatus
Electrical appliances
Insulated wire and cable
Engine electrical equipment
Electric lamps
Radio and related products
Tubes
Communication equipment
Storage batteries
Primary batteries
X-ray apparatus
Motor vehicles
Truck trailers
Automobile trailers
Aircraft and parts
Ships and boats
Locomotives
Railroad equipment
Motorcycles and bicycles
Instruments, etc.
Optical, ophthalmic and photo equipment
Medical and dental instruments and supplies
Watches and clocks
Jewelry and silverware
Musical instruments and parts
Toys and sporting goods
Office supplies
Plastic products
Cork products
Motion picture production
Miscellaneous manufactured products
Electric light and power
Natural, manufactured and mixed gas
Railroads
Trucking
Warehousing and storage
Overseas transportation
Other water transportation
Air transportation
Pipeline transportation
Wholesale trade
Retail trade
Local and highway transportation
Telephone and telegraph
Banking, finance, and insurance
Hotels
Real estate and rentals
Laundries and dry cleaning
Other personal services
Advertising, including radio and television
Business services
Automobile repair services and garages
Other repair services
Motion pictures and other amusements
Medical, dental and other professional services
New construction
Maintenance construction
Waste products, metal
Waste products, nonmetal
Small arms
Small arms ammunition
Iron ore mining
Copper mining
Lead and zinc mining
Bauxite mining
Other mining
Crude petroleum and natural gas
Stone, sand, clay, and abrasives
Other nonmetallic minerals

All Other Industries Sales

Includes the Sales of the Following Industries

Iron ore mining
Copper mining
Lead and zinc mining
Bauxite mining
Other mining
Crude petroleum and natural gas
Stone, sand, clay, and abrasives
Sulfur
Other nonmetallic minerals
Logging
Sawmills, planing, and veneer mills
Plywood
Fabricated wood products
Wood furniture
Metal furniture
Partitions, screens, shades, etc.
Pulp mills
Printing and publishing
Paving and roofing materials
Cement
Structural clay products
Pottery and related products
Concrete and plaster products
Abrasive products
Lighting fixtures
Asbestos products
Other miscellaneous nonmetallic minerals
Blast furnaces
Steel works and rolling mills
Iron foundries
Steel foundries
Primary copper
Copper rolling and drawing
Primary lead
Primary zinc
Primary metals, n.e.c.
Nonferrous metal rolling, n.e.c.
Primary aluminum
Aluminum rolling and drawing
Secondary nonferrous metals
Nonferrous foundries
Iron and steel forgings
Cutlery
Tools and general hardware
Hardware, n.e.c.
Metal plumbing and vitreous fixtures
Heating equipment
Structural metal products
Boiler shop products and pipe bending
Metal coating and engraving
Fabricated wire products
Miscellaneous fabricated metal products
Steel springs
Nuts, bolts, and screw machine products
Steam engines and turbines
Internal combustion engines
Farm and industrial tractors
Farm equipment
Construction and mining machinery
Oil-field machinery and tools
Machine tools and metalworking machinery
Cutting tools, jigs, and fixtures
Special industrial machinery
Pumps and compressors
Elevators and conveyors
Blowers and fans
Power transmission equipment
Industrial machinery, n.e.c.
Commercial machines and equipment, n.e.c.
Refrigeration equipment
Valves and fittings
Ball and roller bearings
Machine shops
Wiring devices and graphite products
Electrical measuring instruments
Motors and generators
Transformers
Electrical control apparatus
Electrical welding apparatus
Electrical appliances
Insulated wire and cable
Engine electrical equipment
Electric lamps
Radio and related products
Tubes
Communication equipment
Storage batteries
Primary batteries
X-ray apparatus
Motor vehicles
Truck trailers
Automobile trailers
Aircraft and parts
Ships and boats
Locomotives
Railroad equipment
Motorcycles and bicycles
Instruments, etc.
Optical, ophthalmic and photo equipment
Medical and dental instruments and supplies
Watches and clocks
Jewelry and silverware
Musical instruments and parts
Toys and sporting goods

n.e.c. = not elsewhere classified

Office supplies
Plastic products
Motion picture production
Miscellaneous manufactured products
Local and highway transportation
Telephone and telegraph
Banking, finance, and insurance
Hotels
Laundries and dry cleaning
Other personal services
Business services
Motion pictures and other amusements
Nonprofit institutions
Small arms
Small arms ammunition
New construction
Waste products, metal
Waste products, nonmetal

APPENDIX II

Tables

TABLE A-1. OPERATING DATA FOR SUPPLIERS OF AGRICULTURAL PRODUCTS: 1947, 1954
(Millions of dollars)

	1947				1954			
	Wholesale* Corporations	Retail* Corporations	Manufacturing Corporations	Total	Wholesale* Corporations	Retail* Corporations	Manufacturing Corporations	Total
Net Sales	$832.1	$4,806.5	$7,301.5	$12,940.1	$1,152.5	$6,523.0	$12,437.5	$20,113.0
Cost of Goods and Expense	797.4	4,589.0	6,408.3	11,794.7	1,101.0	6,276.6	11,102.1	18,479.7
Net Profit from Operations	34.7	217.5	893.2	1,145.4	51.5	246.4	1,335.4	1,633.3
Other Income or Deductions (Net)	4.5	32.9	22.8	60.2	6.3	35.9	86.6	128.8
Net Profit before Fed. Income Taxes	39.2	250.4	916.0	1,205.6	57.8	282.3	1,422.0	1,762.1
Provisions for Fed. Income Taxes	15.3	91.0	323.6	429.9	27.4	126.4	546.8	700.6
Net Profit after Taxes	$23.9	$159.4	$592.4	$775.7	$30.4	$155.9	$875.2	$1,061.5

* Estimated on the basis of retail and wholesale sales indexes as published in *Business Statistics, 1955*, Biennial Edition, U.S. Department of Commerce, Office of Business Economics.

SOURCE: Federal Trade Commission and Securities and Exchange Commission, *U.S. Retail and Wholesale Corporations*, Quarterly Financial Reports; ibid., *U.S. Manufacturing Corporations.*

TABLE A-2. OPERATING DATA FOR PROCESSING AND DISTRIBUTING INDUSTRIES: 1947, 1954
(Millions of dollars)

	1947				1954			
	Wholesale* Corporations	Retail* Corporations	Manufacturing Corporations	Total	Wholesale* Corporations	Retail* Corporations	Manufacturing Corporations	Total
Net Sales	$12,928	$17,168	$48,957	$79,053	$16,819	$24,880	$67,433	$109,132
Cost of Goods and Expense	12,722	16,721	44,844	74,287	16,662	24,358	64,593	105,613
Net Profit from Operations	206	447	4,113	4,766	157	522	2,840	3,519
Other Income or Deductions (Net)	44	66	—6	104	56	101	—76	81
Net Profit before Fed. Income Taxes	250	513	4,107	4,870	213	623	2,764	3,600
Provisions for Fed. Income Taxes	101	227	1,670	1,998	118	317	1,487	1,922
Net Profit after Taxes	$149	$286	$2,437	$2,872	$95	$306	$1,277	$1,678

* Estimated on the basis of retail and wholesale sales indexes as published in *Business Statistics, 1955*, Biennial Edition, U.S. Department of Commerce, Office of Business Economics.

SOURCE: Federal Trade Commission and Securities and Exchange Commission, *U.S. Retail and Wholesale Corporations*, Quarterly Financial Reports; ibid., *U.S. Manufacturing Corporations.*

TABLE A-3. DISTRIBUTION OF NUMBER OF FARMS AND VALUE OF PRODUCTS BY ECONOMIC CLASS: 1949

Economic Class in Terms of Value of Products Sold	*Number of Farms*	*% of Total Number of Farms*	*% of Total Value of Product Sold*
COMMERCIAL FARMS:			
Large-Scale Farms ($25,000 or more)	103,231	1.9%	26.0%
Family-Scale Farms:			
Large ($10,000–$24,999)	381,151	7.1	24.8
Upper Medium ($5,000–$9,999)	721,211	13.4	22.7
Lower Medium ($2,500–$4,999)	882,302	16.4	14.4
Small ($1,200–$2,499)	901,316	16.8	7.3
Small-Scale Farms ($250–$1,199) *	717,201	13.3	2.3
TOTAL COMMERCIAL FARMS	3,706,412	68.9%	97.5%
NONCOMMERCIAL FARMS:			
Part-Time ($250–$1,199) †	639,230	11.9	1.7
Residential (less than $250)	1,029,392	19.1	0.4
Unusual Institutional Farms, etc.	4,216	0.1	0.4
ALL NONCOMMERCIAL FARMS	1,672,838	31.1%	2.5%
ALL FARMS	5,379,250	100.0%	100.0%

* Less than 100 days of off-farm work by operator and income of operator and members of his family from nonfarm sources less than value of all farm products sold.

† One hundred days or more of off-farm work by opertaor and income of operator and members of his family from nonfarm sources more than value of all farm products sold.

SOURCE: U.S. Census of Agriculture, 1950.

TABLE A-4. INDUSTRIAL RESEARCH AND DEVELOPMENT EXPENSE: 1951

	Total Research		
	Dollars (Millions)	*% of Total Sales*	*% Financed by Federal Government*
All Industries	$1,783.7	2.0%	47.0%
Manufacturing	1,613.5	2.0	46.5
Food and Kindred Products	23.8	0.3	3.7
Textile Mill Products and Apparel	15.8	0.9	14.4
Machinery (except Electrical)	99.3	1.5	23.9

SOURCE: *Statistical Abstract of the United States*, 1955, Bureau of the Census, p. 499.

TABLE A-5. ESTIMATED RESEARCH AND DEVELOPMENT EXPENDITURES ON AGRICULTURAL PROBLEMS BY SELECTED INDUSTRIAL CONCERNS: 1951*
(Millions of dollars)

Type of Research	*Amount*
Agricultural Production Research	
Farm Machinery	$20
Agricultural Chemicals, Fertilizers, Insecticides, Feeds, etc.	25
Other	5
Subtotal	$50
Utilization Research on Agricultural Products	
Food and Related Products	41
Forest Products	36
Agricultural Textiles, Tobacco, and Miscellaneous Products	13
Subtotal	$90
Total	$140

* The above estimate includes data from some 300 companies which carry on research relating to agriculture. It is based on figures supplied by the Department of Defense, Department of Labor, National Research Council, and certain data collected by the authors. It excludes expenditures relating to quality control, product testing, market research, sales promotion, sales service and research in the social sciences and psychology.

The figures represent a conservative estimate since they do not include expenditures of approximately 130 companies known to be doing some research on agricultural problems. No adequate data were available from these 130 companies.

Prepared in ARS-BFD, U.S.D.A. Budgetary Reports, November 5, 1954.

TABLE A-6. EXPENDITURES OF THE FEDERAL GOVERNMENT FOR SCIENTIFIC RESEARCH AND DEVELOPMENT: 1947, 1954, AND 1956*
(Millions of dollars)

	1947	*1954*	*1956†*
Department of Agriculture	$39.2	$55.5	$80.8
Department of Commerce	16.8	20.1	37.5
Department of Defense	529.2	1,532.2	1,520.4
Department of Health, Education, and Welfare	12.5	61.6	76.4
Department of the Interior	24.9	40.3	36.1
Atomic Energy Commission	52.5	274.3	344.0
Manhattan Engineer District	186.0	—	—
National Advisory Committee for Aeronautics	33.5	89.5	76.0
Office of Scientific Research and Development	5.6	—	—
All Other	15.4	29.0	57.9
Total, All Agencies	$915.6	$2,102.5	$2,229.1

* In addition to Federal Government expenditures, State Agricultural Experiment Stations spent (in addition to Federal grants to them) $40.3 million in 1947, $67.2 million in 1954, and $72.2 million in 1955 for agricultural research.

† Estimated.

SOURCE: *The Federal Research and Development Budget*, National Science Foundation, Office of Special Studies, Publication 4 (Washington, Government Printing Office, 1955), p. 38.

TABLE A-7. SELECTED EXPENDITURES FOR RESEARCH BY 37 FOUNDATIONS REPORTING FOR THE YEARS 1946 AND 1953

	1946		1953	
Type of Research	*Dollars (Millions)*	*% of Total*	*Dollars (Millions)*	*% of Total*
Total Expenditures	$44.5	100.0%	$117.5	100.0%
Scientific Research	$10.9	24.5%	$23.2	19.8%
Life Sciences	5.7	12.9	10.7	9.1
Medical	4.0	9.1	6.5	5.6
Agricultural	0.3	0.6	0.7	0.6
Biological	1.4	3.2	3.4	2.9
Physical Sciences	2.5	5.5	1.9	1.6
Social Sciences	2.7	6.1	10.7	9.1

SOURCE: *Scientific Research Expenditures by the Larger Private Foundations* (National Science Foundation, 1956), p. 15. Appendix C.

TABLE A-8. SOAP AND RELATED PRODUCTS SECTOR PURCHASES: 1947

Sellers	*Rank within Sector*	*Gross Purchases* Dollars (Millions)	% of Total
Labor and Capital Utilization	1	$388	25.3%
Animal Oils	2	374	24.4
Other Chemical Industries	3	220	14.3
Vegetable Oils	4	150	9.8
Advertising	5	122	7.9
Government Services (Taxes, etc.)	6	69	4.5
Paper and Board Mills and Converted Paper Products	7	66	4.3
Railroads	8	23	1.5
Wholesale Trade	9	9	0.6
Metal Container Materials and Cork Products	10	8	0.5
Tin Cans and Other Tins	11	7	0.5
Petroleum Products	11	7	0.5
Tires, Tubes, and Other Rubber Products	13	6	0.4
Glass	14	5	0.3
Meat Animals and Products	15	4	0.3
Other Fuels (Coal, Coke, and Gas)	15	4	0.3
Trucking	15	4	0.3
All Other Agriculture	18	2	0.1
Miscellaneous Food Products	18	2	0.1
Automobile and Other Repair Services	18	2	0.1
Water and Other Transportation	18	2	0.1
Real Estate and Rentals	18	2	0.1
Maintenance and Construction	18	2	0.1
Electric Light and Power	18	2	0.1
Meat Packing and Wholesale Poultry	25	1	0.1
Canning, Preserving, and Freezing	25	1	0.1
Grain Mill Products	25	1	0.1
Wood Containers and Cooperage	25	1	0.1
Fertilizers	25	1	0.1
Processed Dairy Products		†	**
Bakery Products		†	**
Sugar		†	**
Alcoholic Beverages		†	**
Spinning, Weaving, and Dyeing		†	**
Warehouse and Storage		†	**
Retail Trade		†	**
All Other Industries	—	48	3.1
Total	—	$1,533	100.0%

** Less than 0.05%; note that these amounts are not included in the total.

† Sales of less than $500,000; note that these amounts are not included in the total.

SOURCE: Derived from data published by the Bureau of Labor Statistics, Division of Interindustry Economics.

TABLE A-9. PAINTS AND ALLIED PRODUCTS SECTOR PURCHASES: 1947

Sellers	Rank within Sector	Gross Purchases Dollars (Millions)	% of Total
Labor and Capital Utilization	1	$521	32.1%
Other Chemical Industries	2	431	26.5
Vegetable Oils	3	188	11.6
Government Services (Taxes, etc.)	4	70	4.3
Petroleum Products	5	37	2.3
Tin Cans and Other Tins	6	34	2.1
Railroads	7	28	1.7
Paper and Board Mills and Converted Paper Products	8	19	1.2
Metal Container Materials and Cork Products	9	18	1.1
All Other Agriculture	10	15	0.9
Wholesale Trade	11	14	0.9
Advertising	12	12	0.7
Other Fuels (Coal, Coke, and Gas)	13	10	0.6
Animal Oils	14	8	0.5
Electric Light and Power	15	6	0.4
Tires, Tubes, and Other Rubber Products	16	5	0.3
Real Estate and Rentals	16	5	0.3
Trucking	18	4	0.2
Water and Other Transportation	19	3	0.2
Processed Dairy Products	20	2	0.1
Automobile and Other Repair Services	20	2	0.1
Maintenance and Construction	20	2	0.1
Miscellaneous Food Products	23	1	0.1
Glass	23	1	0.1
Warehouse and Storage	23	1	0.1
Meat Packing and Wholesale Poultry		†	**
Canning, Preserving, and Freezing		†	**
Grain Mill Products		†	**
Bakery Products		†	**
Sugar		†	**
Alcoholic Beverages		†	**
Spinning, Weaving, and Dyeing		†	**
Wood Containers and Cooperage		†	**
Fertilizers		†	**
Retail Trade		†	**
All Other Industries	—	186	11.5
Total	—	$1,623	100.0%

** Less than 0.05%; note that these amounts are not included in the total.

† Sales of less than $500,000; note that these amounts are not included in the total.

SOURCE: Derived from data published by the Bureau of Labor Statistics, Division of Interindustry Economics.

TABLE A-10. GUM AND WOOD CHEMICALS SECTOR PURCHASES: 1947

Sellers	*Rank within Sector*	*Gross Purchases* Dollars (Millions)	% of Total
Labor and Capital Utilization	1	$66	42.3%
All Other Agriculture	2	39	25.0
Other Chemical Industries	3	8	5.2
Government Services (Taxes, etc.)	4	7	4.5
Metal Container Materials and Cork Products	5	4	2.6
Railroads	6	3	1.9
Trucking	6	3	1.9
Other Fuels (Coal, Coke, and Gas)	8	2	1.3
Tires, Tubes, and Other Rubber Products	9	1	0.6
Wholesale Trade	9	1	0.6
Real Estate and Rentals	9	1	0.6
Wood Containers and Cooperage	9	1	0.6
Meat Packing and Wholesale Poultry		†	**
Canning, Preserving, and Freezing		†	**
Grain Mill Products		†	**
Bakery Products		†	**
Miscellaneous Food Products		†	**
Sugar		†	**
Alcoholic Beverages		†	**
Vegetable Oils		†	**
Animal Oils		†	**
Paper and Board Mills and Converted Paper Products		†	**
Glass		†	**
Fertilizers		†	**
Petroleum Products		†	**
Electric Light and Power		†	**
Automobile and Other Repair Services		†	**
Water and Other Transportation		†	**
Warehouse and Storage		†	**
Retail Trade		†	**
Advertising		†	**
Maintenance and Construction		†	**
Processed Dairy Products		†	**
All Other Industries	—	20	12.9
Total	—	$156	100.0%

** Less than 0.05%; note that these amounts are not included in the total.

† Sales of less than $500,000; note that these amounts are not included in the total.

SOURCE: Derived from data published by the Bureau of Labor Statistics, Division of Interindustry Economics.

TABLE A-11. SAW MILLS, PLANING, AND VENEER MILLS SECTOR PURCHASES: 1947

Sellers	*Rank within Sector*	*Gross Purchases* Dollars (Millions)	% of Total
Labor and Capital Utilization	1	$1,582	49.4%
Government Services (Taxes, etc.)	2	225	7.0
All Other Agriculture	3	184	5.7
Railroads	4	72	2.2
Trucking	5	63	2.0
Other Fuels (Coal, Coke, and Gas)	6	38	1.2
Automobile and Other Repair Services	7	33	1.0
Petroleum Products	8	30	0.9
Wholesale Trade	9	24	0.7
Tires, Tubes, and Other Rubber Products	10	13	0.4
Other Chemical Industries	11	12	0.4
Electric Light and Power	12	10	0.3
Leather and Leather Goods	13	9	0.3
Wood Containers and Cooperage	14	8	0.3
Advertising	14	8	0.3
Real Estate and Rentals	14	8	0.3
Water and Other Transportation	17	6	0.2
Maintenance and Construction	18	5	0.2
Food Grains and Feed Crops	19	4	0.1
Jute, Linens, Cord, and Twine	20	1	*
Retail Trade	20	1	*
Spinning, Weaving, and Dyeing		†	**
Paper and Board Mills and Converted Paper Products		†	**
Warehouse and Storage		†	**
All Other Industries	—	864	27.0
Total	—	$3,200	100.0%

* Less than 0.05%.

** Less than 0.05%; note that these amounts are not included in the total.

† Sales of less than $500,000; note that these amounts are not included in the total.

SOURCE: Derived from data published by the Bureau of Labor Statistics, Division of Interindustry Economics.

TABLE A-12. EATING AND DRINKING PLACES SECTOR PURCHASES: 1947

Sellers	*Rank within Sector*	*Gross Purchases* Dollars (Millions)	% of Total
Labor and Capital Utilization	1	$4,625	34.9%
Government Services (Taxes, etc.)	2	1,413	10.7
Wholesale Trade	3	1,050	7.9
Alcoholic Beverages	4	986	7.4
Meat Packing and Wholesale Poultry	5	762	5.8
Processed Dairy Products	6	586	4.4
Real Estate and Rentals	7	386	2.9
Miscellaneous Food Products	8	385	2.9
Canning, Preserving, and Freezing	9	383	2.9
Poultry and Eggs	10	358	2.7
Bakery Products	11	267	2.0
Vegetables and Fruits	12	262	2.0
Railroads	13	254	1.9
Farm Dairy Products	14	165	1.2
Other Fuels (Coal, Coke, and Gas)	15	121	0.9
Electric Light and Power	16	99	0.7
Grain Mill Products	17	95	0.7
Automobile and Other Repair Services	18	85	0.6
Trucking	19	82	0.6
Fishing, Hunting, and Trapping	20	73	0.6
Maintenance and Construction	20	73	0.6
Sugar	22	71	0.5
Paper and Board Mills and Converted Paper Products	23	61	0.5
Advertising	24	55	0.4
Other Chemical Industries	25	44	0.3
Metal Container Materials and Cork Products	26	25	0.2
House Furnishings and Other Nonapparel	27	19	0.1
Petroleum Products	28	17	0.1
Glass	29	15	0.1
Retail Trade	30	11	0.1
Warehouse and Storage	31	9	0.1
All Other Agriculture	32	7	0.1
Water and Other Transportation	33	5	*
Tires, Tubes, and Other Rubber Products	34	4	*
Apparel	35	2	*
All Other Industries	—	412	3.1
Total	—	$13,267	100.0%

* Less than 0.05%.

SOURCE: Derived from data published by the Bureau of Labor Statistics, Division of Interindustry Economics.

TABLE A-13. MEDICAL, DENTAL, AND OTHER PROFESSIONAL SERVICES SECTOR PURCHASES: 1947

Sellers	*Rank within Sector*	*Gross Purchases* Dollars (Millions)	*Gross Purchases* % of Total
Labor and Capital Utilization	1	$6,832	76.4%
Real Estate and Rentals	2	263	2.9
Other Chemical Industries	3	232	2.6
Wholesale Trade	4	216	2.4
Government Services (Taxes, etc.)	5	167	1.9
Meat Packing and Wholesale Poultry	6	104	1.2
Maintenance and Construction	7	79	0.9
Farm Dairy Products	8	41	0.4
Miscellaneous Food Products	9	40	0.4
Poultry and Eggs	10	38	0.4
Processed Dairy Products	11	37	0.4
Railroads	12	33	0.4
Petroleum Products	13	30	0.3
Retail Trade	14	28	0.3
Vegetables and Fruits	15	27	0.3
Bakery Products	15	27	0.3
Automobile and Other Repair Services	17	26	0.3
Canning, Preserving, and Freezing	18	25	0.3
Other Fuels (Coal, Coke, and Gas)	18	25	0.3
Electric Light and Power	20	24	0.3
Apparel	21	16	0.2
Paper and Board Mills and Converted Paper Products	21	16	0.2
Trucking	23	11	0.1
Advertising	23	11	0.1
Water and Other Transportation	25	10	0.1
Tires, Tubes, and Other Rubber Products	26	8	0.1
Fishing, Hunting, and Trapping	27	6	0.1
Grain Mill Products	27	6	0.1
Leather and Leather Goods	27	6	0.1
Sugar	30	5	*
Spinning, Weaving, and Dyeing	31	3	*
Glass	31	3	*
House Furnishings and Other Nonapparel	33	2	*
Metal Container Materials and Cork Products	33	2	*
All Other Agriculture	35	1	*
Warehouse and Storage	35	1	*
Meat Animals and Products		†	**
Special Textile Products		†	**
Wood Containers and Cooperage		†	**
Fertilizers		†	**
All Other Industries	—	546	6.1
Total	—	$8,947	100.0%

* Less than 0.05%.

** Less than 0.05%; note that these amounts are not included in the total.

† Sales of less than $500,000; note that these amounts are not included in the total.

SOURCE: Derived from data published by the Bureau of Labor Statistics, Division of Interindustry Economics.

TABLE A-14. ALL OTHER INDUSTRIES AGGREGATE PURCHASES: 1947

Sellers	*Rank within Aggregate*	*Gross Purchases* Dollars (Millions)	% of Total
Labor and Capital Utilization	1	$143,849	48.6%
Government Services	2	20,821	7.0
Maintenance and Construction	3	6,362	2.2
Other Fuels (Coal, Coke, and Gas)	4	5,118	1.7
Paper and Board Mills and Converted Paper Products	5	4,918	1.7
Other Chemical Industries	6	4,912	1.7
Real Estate and Rentals	7	4,337	1.5
Railroads	8	4,280	1.4
Wholesale Trade	9	4,164	1.4
Electric Light and Power	10	3,701	1.3
Petroleum Products	11	3,653	1.2
Advertising	12	2,750	0.9
Automobile and Other Repair Services	13	2,217	0.7
Retail Trade	14	2,183	0.7
Trucking	15	1,613	0.5
Tires, Tubes, and Other Rubber Products	16	1,374	0.5
Water and Other Transportation	17	1,232	0.4
Metal Container Materials and Cork Products	18	1,209	0.4
Glass	19	589	0.2
Spinning, Weaving, and Dyeing	20	509	0.2
Imports (Noncompetitive)	21	432	0.1
House Furnishings and Other Nonapparel	22	336	0.1
Wood Containers and Cooperage	23	270	0.1
Miscellaneous Food Products	24	233	0.1
Special Textile Products	25	175	0.1
Warehouse and Storage	26	174	0.1
Leather and Leather Goods	27	167	0.1
All Other Agriculture	28	147	*
Vegetable Oils	29	132	*
Animal Oils	30	105	*
Tin Cans and Other Tins	31	95	*
Meat Packing and Wholesale Poultry	32	85	*
Fertilizers	33	84	*
Jute, Linens, Cord, and Twine	34	83	*
Processed Dairy Products	35	57	*
Apparel	36	56	*
Sugar	37	42	*
Food Grains and Feed Crops	38	38	*
Alcoholic Beverages	38	38	*
Canvas Products	40	29	*
Bakery Products	41	15	*
Grain Mill Products	42	14	*
Canning, Preserving, and Freezing	43	12	*
Tobacco Manufactures	44	10	*
Vegetables and Fruits	45	9	*
Fishing, Hunting, and Trapping	46	5	*
Cotton	47	4	*
Farm Dairy Products	48	3	*
Poultry and Eggs	48	3	*
Meat Animals and Products	50	1	*
All Other Industries	—	73,118	24.7
Total	—	$295,763	100.0%

* Less than 0.05%.

SOURCE: Derived from data published by the Bureau of Labor Statistics, Division of Interindustry Economics.

TABLE A-15. EXPORTS (MINUS COMPETITIVE IMPORTS) SECTOR PURCHASES: 1947

Sellers	Rank within Sector	Gross Purchases Dollars (Millions)	% of Total
Water and Other Transportation	1	$1,691	12.8%
Government Services	2	1,251	9.5
Wholesale Trade	3	1,037	7.9
Food Grains and Feed Crops	4	919	7.0
Labor and Capital Utilization	5	914	6.9
Spinning, Weaving, and Dyeing	6	836	6.3
Grain Mill Products	7	757	5.7
Railroads	8	683	5.2
Petroleum Products	9	427	3.2
Other Fuels (Coal, Coke, and Gas)	10	375	2.8
Cotton	11	344	2.6
Meat Packing and Wholesale Poultry	12	271	2.1
Other Chemical Industries	13	240	1.8
Processed Dairy Products	14	239	1.8
Apparel	15	180	1.4
Tires, Tubes, and Other Rubber Products	16	168	1.3
Trucking	17	150	1.1
Miscellaneous Food Products	18	121	0.9
Tobacco Manufactures	19	89	0.7
Tobacco	20	88	0.7
House Furnishings and Other Nonapparel	20	88	0.7
Glass	22	66	0.5
Leather and Leather Goods	23	47	0.4
Metal Container Materials and Cork Products	24	44	0.3
Warehouse and Storage	25	26	0.2
Poultry and Eggs	26	16	0.1
Vegetables and Fruits	26	16	0.1
Tin Cans and Other Tins	28	13	0.1
Bakery Products	29	8	0.1
Animal Oils	30	5	*
Wood Containers and Cooperage	30	5	*
Fertilizers	30	5	*
Canvas Products	33	3	*
Advertising	33	3	*
Retail Trade	35	2	*
Farm Dairy Products	36	1	*
All Other Industries	—	4,032	30.6
Electric Light and Power	37	— 4	*
Jute, Linens, Cord, and Twine	38	— 16	—0.1
Special Textile Products	39	— 44	—0.3
Vegetable Oils	40	— 56	—0.4
Alcoholic Beverages	41	— 63	—0.5
All Other Agriculture	42	—102	—0.8
Canning, Preserving, and Freezing	43	—106	—0.8
Oil-Bearing Crops	44	—115	—0.9
Fishing, Hunting, and Trapping	45	—147	—1.1
Paper and Board Mills and Converted Paper Products	46	—236	—1.8
Meat Animals and Products	47	—324	—2.5
Sugar	48	—765	—5.8
Total	—	$13,182	100.0%

* Less than 0.05%.

SOURCE: Derived from data published by the Bureau of Labor Statistics, Division of Interindustry Economics.

TABLE A-16. GROSS PRIVATE CAPITAL FORMATION SECTOR PURCHASES: 1947

Sellers	*Rank within Sector*	*Gross Purchases*	
		Dollars (Millions)	*% of Total*
Wholesale Trade	1	$1,379	4.1%
Retail Trade	2	1,049	3.1
Real Estate and Rentals	3	804	2.4
Railroads	4	268	0.8
Labor and Capital Utilization	5	232	0.7
Government Services (Taxes, etc.)	6	216	0.6
Trucking	7	93	0.3
Metal Container Materials and Cork Products	8	24	0.1
Glass	9	21	0.1
Meat Animals and Meat Products	9	21	0.1
Special Textile Products	9	21	0.1
Leather and Leather Goods	12	17	*
Wood Containers and Cooperage	13	13	*
Water and Other Transportation	14	9	*
Tires, Tubes, and Other Rubber Products	15	7	*
Warehouse and Storage	16	2	*
House Furnishings and Other Nonapparel	17	1	*
Spinning, Weaving, and Dyeing		†	**
Jute, Linens, Cord, and Twine		†	**
Canvas Products		†	**
Petroleum Products		†	**
All Other Industries	—	29,487	87.5
Total	—	$33,664	100.0%

* Less than 0.05%.

** Less than 0.05%; note that these amounts are not included in the total.

† Sales of less than $500,000; note that these amounts are not included in the total.

SOURCE: Derived from data published by the Bureau of Labor Statistics, Division of Interindustry Economics.

TABLE A-17. INVENTORY CHANGES BY SECTORS: 1947

Sector	Rank	Dollars (Millions)	% of Total
Other Chemical Industries	1	+$283	+17.2%
Apparel	2	+ 180	+11.0
Leather and Leather Goods	3	+ 166	+10.1
Oil-Bearing Crops	4	+ 143	+ 8.7
Petroleum Products	5	+ 134	+ 8.2
Meat Packing and Wholesale Poultry	6	+ 120	+ 7.3
Wholesale Trade	7	+ 113	+ 6.9
Tobacco Manufactures	8	+ 109	+ 6.6
Canning, Preserving, and Freezing	9	+ 99	+ 6.0
Tires, Tubes, and Other Rubber Products	10	+ 87	+ 5.3
Spinning, Weaving, and Dyeing	11	+ 85	+ 5.2
Railroads	12	+ 70	+ 4.3
Government Services (Taxes, etc.)	13	+ 65	+ 4.0
Alcoholic Beverages	14	+ 61	+ 3.7
Metal Container Materials and Cork Products	15	+ 46	+ 2.8
Cotton	16	+ 35	+ 2.1
Animal Oils	17	+ 28	+ 1.7
Glass	17	+ 28	+ 1.7
Trucking	19	+ 27	+ 1.6
Tobacco	20	+ 19	+ 1.2
Other Fuels (Coal, Coke, and Gas)	21	+ 18	+ 1.1
Vegetable Oils	22	+ 16	+ 1.0
Paper and Board Mills and Converted Paper Products	23	+ 15	+ 0.9
Special Textile Products	24	+ 14	+ 0.9
Jute, Linens, Cord, and Twine	25	+ 12	+ 0.7
Sugar	26	+ 10	+ 0.6
Tin Cans and Other Tins	27	+ 6	+ 0.4
Wood Containers and Cooperage	28	+ 5	+ 0.3
House Furnishings and Other Nonapparel	29	+ 4	+ 0.2
Processed Dairy Products	29	+ 4	+ 0.2
Water and Other Transportation	29	+ 4	+ 0.2
Warehouse and Storage	32	+ 3	+ 0.2
Fertilizers	33	+ 2	+ 0.1
Miscellaneous Food Products	34	+ 1	+ 0.1
Farm Dairy Products	35	†	**
All Other Industries	—	+1,505	+91.6
Canvas Products	36	— 1	— 0.1
All Other Agriculture	37	— 11	— 0.7
Imports (Noncompetitive)	38	— 20	— 1.2
Bakery Products	39	— 21	— 1.3
Poultry and Eggs	40	— 43	— 2.6
Vegetables and Fruits	41	— 96	— 5.8
Grain Mill Products	42	— 99	— 6.0
Meat Animals and Products	43	— 680	—41.4
Food Grains and Feed Crops	44	— 903	—55.0
Total	—	+$1,643	100.0%

† Less than $500,000; note that this amount is not included in the total.

** Less than 0.05%; note that this amount is not included in the total.

SOURCE: Derived from data published by the Bureau of Labor Statistics, Division of Interindustry Economics.

TABLE A-18. GROSS DOMESTIC PURCHASES BY SECTORS: 1947

Buyers	Rank within the National Economy	Gross Purchases: Dollars (Millions)	% of Respective Aggregate†	% of Total National Economy
Consumer Purchases (Households)	1	$197,537		27.0%
Government Purchases	2	51,151		7.0
Food Processing Aggregate	3	41,970		5.7
Meat Packing and Wholesale Poultry		11,106	26.5%	1.5
Miscellaneous Food Products		6,634	15.8	0.9
Grain Mill Products		5,345	12.7	0.7
Processed Dairy Products		3,646	8.7	0.5
Bakery Products		3,351	8.0	0.4
Canning, Preserving, and Freezing		2,728	6.5	0.4
Alcoholic Beverages		2,723	6.5	0.4
Tobacco Manufactures		2,567	6.1	0.3
Vegetable Oils		1,913	4.5	0.3
Sugar		1,180	2.8	0.2
Animal Oils		777	1.9	0.1
Farming Aggregate	4	40,277		5.5
Food Grains and Feed Crops		11,006	27.3	1.5
Meat Animals and Products		9,803	24.3	1.3
Farm Dairy Products		5,066	12.6	0.7
Vegetables and Fruits		4,012	10.0	0.6
Poultry and Eggs		3,862	9.6	0.5
Cotton		2,222	5.5	0.3
All Other Agriculture		1,957	4.9	0.3
Oil-Bearing Crops		1,059	2.6	0.1
Tobacco		885	2.2	0.1
Fishing, Hunting, and Trapping		405	1.0	0.1
Gross Private Capital Formation	5	33,664		4.6
Fiber Processing Aggregate	6	26,133		3.6
Apparel		11,333	43.4	1.6
Spinning, Weaving, and Dyeing		8,097	31.0	1.1
Leather and Leather Goods		3,724	14.2	0.5
House Furnishings and Other Nonapparel		1,806	6.9	0.2
Special Textile Products		823	3.1	0.1
Jute, Linens, Cord, and Twine		253	1.0	*
Canvas Products		97	0.4	*
Eating and Drinking Places	7	13,267		1.8
Exports (minus Competitive Imports)	8	13,182		1.8
Medical and Other Professional Services	9	8,947		1.2
Saw Mills, Planing, and Veneer Mills	10	3,200		0.4
Tires, Tubes, and Other Rubber Products	11	2,998		0.4
Inventory Changes	12	1,643		0.2
Paints and Allied Products	13	1,623		0.2
Soap and Related Products	14	1,533		0.2
Gum and Wood Chemicals	15	156		*
All Other Industries	—	295,763		40.4
Total Gross Domestic Output	—	$733,044		100.0%

* Less than 0.05%.

† Sectors which are not combined into Aggregates are valued at 100.0%.

SOURCE: Derived from data published by the Bureau of Labor Statistics, Division of Interindustry Economics.

TABLE A-19. CONTAINER AGGREGATE SALES: 1947

Buyers	Rank within Aggregate	Gross Sales: Dollars (Millions)	Gross Sales: % of Total
Consumer Purchases (Households)	1	$725	7.0%
Canning, Preserving, and Freezing	2	402	3.9
Miscellaneous Food Products	3	397	3.8
Alcoholic Beverages	4	259	2.5
Processed Dairy Products	5	156	1.5
Vegetables and Fruits	6	134	1.3
Bakery Products	7	132	1.3
Meat Packing and Wholesale Poultry	8	117	1.1
Tobacco Manufactures	9	116	1.1
Eating and Drinking Places	10	101	1.0
Inventory Changes	11	100	1.0
Grain Mill Products	12	93	0.9
Soap and Related Products	13	87	0.8
Government Purchases	13	87	0.8
Leather and Leather Goods	15	85	0.8
Apparel	16	76	0.7
Paints and Allied Products	17	72	0.7
Gross Private Capital Formation	18	58	0.5
Tires, Tubes, and Other Rubber Products	19	48	0.5
Spinning, Weaving, and Dyeing	20	44	0.4
Farm Dairy Products	21	30	0.3
House Furnishings and Other Nonapparel	22	27	0.2
Medical, Dental, and Other Professional Services	23	21	0.2
Special Textile Products	24	11	0.1
Sugar	25	10	0.1
Saw Mills, Planing, and Veneer Mills	26	8	0.1
Vegetable Oils	27	6	0.1
Gum and Wood Chemicals	28	5	*
Fishing, Hunting, and Trapping	29	3	*
All Other Agriculture	30	2	*
Animal Oils	30	2	*
Jute, Linens, Cord, and Twine	30	2	*
All Other Industries	—	7,081	68.2
Imports (Noncompetitive)	—	—108	—1.0
Total	—	$10,389	100.0%

* Less than 0.05%.

SOURCE: Derived from data published by the Bureau of Labor Statistics, Division of Interindustry Economics.

TABLE A-20. FERTILIZERS SECTOR SALES: 1947

Buyers	*Rank within Sector*	*Gross Sales* *Dollars (Millions)*	*% of Total*
Food Grains and Feed Crops	1	$246	47.2%
Vegetables and Fruits	2	76	14.6
Cotton	3	40	7.7
Tobacco	4	28	5.4
All Other Agriculture	5	13	2.5
Oil-Bearing Crops	6	12	2.3
Consumer Purchases (Households)	7	8	1.5
Exports (minus Competitive Imports)	8	5	0.9
Animal Oils	9	4	0.8
Government Purchases	10	2	0.4
Inventory Changes	10	2	0.4
Soap and Related Products	12	1	0.2
Paints and Allied Products		†	**
Gum and Wood Chemicals		†	**
Medical, Dental, and Other Professional Services		†	**
All Other Industries	—	84	16.1
Total	—	$521	100.0%

** Less than 0.05%; note that these amounts are not included in the total.

† Purchases of less than $500,000; note that these amounts are not included in the total.

SOURCE: Derived from data published by the Bureau of Labor Statistics, Division of Interindustry Economics.

TABLE A-21. POWER AGGREGATE (PETROLEUM, GAS, AND ELECTRIC) SALES: 1947

Buyers	Rank within Aggregate	Gross Sales: Dollars (Millions)	Gross Sales: % of Total
Consumer Purchases (Households)	1	$2,576	14.3%
Exports (minus Competitive Imports)	2	798	4.4
Government Purchases	3	436	2.4
Food Grains and Feed Crops	4	245	1.4
Eating and Drinking Places	5	237	1.3
Inventory Changes	6	152	0.8
Spinning, Weaving, and Dyeing	7	122	0.7
Medical, Dental, and Other Professional Services	8	79	0.4
Saw Mills, Planing, and Veneer Mills	9	78	0.4
Miscellaneous Food Products	10	70	0.4
Vegetables and Fruits	11	57	0.3
Paints and Allied Products	12	53	0.3
Tires, Tubes, and Other Rubber Products	13	52	0.3
All Other Agriculture	14	51	0.3
Farm Dairy Products	15	48	0.3
Meat Packing and Wholesale Poultry	16	43	0.2
Bakery Products	17	42	0.2
Processed Dairy Products	18	36	0.2
Apparel	18	36	0.2
Meat Animals and Products	20	32	0.2
Grain Mill Products	20	32	0.2
Cotton	22	30	0.2
Alcoholic Beverages	23	27	0.2
Poultry and Eggs	24	25	0.1
Oil-Bearing Crops	25	21	0.1
Canning, Preserving, and Freezing	26	18	0.1
Sugar	27	17	0.1
Leather and Leather Goods	27	17	0.1
Soap and Related Products	29	13	0.1
Vegetable Oils	30	11	0.1
House Furnishings and Other Nonapparel	31	9	0.1
Special Textile Products	32	8	*
Fishing, Hunting, and Trapping	33	7	*
Tobacco	34	5	*
Animal Oils	34	5	*
Tobacco Manufactures	36	4	*
Jute, Linens, Cord, and Twine	37	2	*
Gum and Wood Chemicals	37	2	*
Canvas Products		†	**
Gross Private Capital Formation		†	**
All Other Industries	—	12,472	69.4
Total	—	$17,968	100.0%

* Less than 0.05%.

** Less than 0.05%; note that these amounts are not included in the total.

† Purchases of less than $500,000; note that these amounts are not included in the total.

SOURCE: Derived from data published by the Bureau of Labor Statistics, Division of Interindustry Economics.

TABLE A-22. OTHER CHEMICAL INDUSTRIES SECTOR SALES: 1947

Buyers	Rank within Sector	Gross Sales Dollars (Millions)	% of Total
Consumer Purchases (Households)	1	$1,988	18.5%
Spinning, Weaving, and Dyeing	2	685	6.4
Tires, Tubes, and Other Rubber Products	3	621	5.8
Paints and Allied Products	4	431	4.0
Inventory Changes	5	283	2.6
Exports (minus Competitive Imports)	6	240	2.2
Medical, Dental, and Other Professional Services	7	232	2.2
Soap and Related Products	8	220	2.1
Tobacco	9	134	1.2
Government Purchases	10	126	1.2
Leather and Leather Goods	11	122	1.1
Miscellaneous Food Products	12	104	1.0
Animal Oils	13	78	0.7
Vegetables and Fruits	14	65	0.6
House Furnishings and Other Nonapparel	15	58	0.5
Food Grains and Feed Crops	16	52	0.5
Eating and Drinking Places	17	44	0.4
Grain Mill Products	18	42	0.4
Farm Dairy Products	19	41	0.4
Meat Packing and Wholesale Poultry	20	37	0.3
Meat Animals and Products	21	28	0.3
Tobacco Manufactures	22	25	0.2
Alcoholic Beverages	23	23	0.2
Bakery Products	24	21	0.2
Processed Dairy Products	25	20	0.2
Cotton	26	17	0.2
Vegetable Oils	27	14	0.1
Canning, Preserving, and Freezing	28	12	0.1
Saw Mills, Planing, and Veneer Mills	28	12	0.1
Gum and Wood Chemicals	30	8	0.1
Tobacco	31	6	0.1
Special Textile Products	31	6	0.1
Poultry and Eggs	33	5	0.1
All Other Agriculture	34	4	*
Sugar	34	4	*
Oil-Bearing Crops	36	2	*
Jute, Linens, Cord, and Twine	36	2	*
Miscellaneous Food Products	38	1	*
Canvas Products	38	1	*
All Other Industries	—	4,912	45.8
Total	—	$10,726	100.0%

* Less than 0.05%.

SOURCE: Derived from data published by the Bureau of Labor Statistics, Division of Interindustry Economics.

TABLE A-23. AUTOMOBILE AND OTHER REPAIR SERVICES SECTOR SALES: 1947

Buyers	Rank within Sector	Gross Sales Dollars (Millions)	% of Total
Consumer Purchases (Households)	1	$2,585	47.0%
Food Grains and Feed Crops	2	158	2.9
Eating and Drinking Places	3	85	1.5
Government Purchases	4	57	1.0
Bakery Products	5	51	0.9
Vegetables and Fruits	6	38	0.7
Miscellaneous Food Products	7	35	0.6
All Other Agriculture	8	33	0.6
Saw Mills, Planing, and Veneer Mills	8	33	0.6
Medical, Dental, and Other Professional Services	10	26	0.5
Meat Animals and Products	11	25	0.5
Meat Packing and Wholesale Poultry	12	19	0.3
Processed Dairy Products	13	18	0.3
Farm Dairy Products	14	17	0.3
Cotton	15	16	0.3
Oil-Bearing Crops	16	14	0.3
Spinning, Weaving, and Dyeing	17	11	0.2
Canning, Preserving, and Freezing	18	9	0.2
Grain Mill Products	18	9	0.2
Poultry and Eggs	20	8	0.1
Alcoholic Beverages	20	8	0.1
Tires, Tubes, and Other Rubber Products	22	5	0.1
Apparel	23	4	0.1
Tobacco	24	3	0.1
Animal Oils	24	3	0.1
Vegetable Oils	26	2	*
Special Textile Products	26	2	*
Soap and Related Products	26	2	*
Paints and Allied Products	26	2	*
Sugar	30	1	*
Canvas Products	30	1	*
House Furnishings and Other Nonapparel	30	1	*
Leather and Leather Goods	30	1	*
Tobacco Manufactures	30	1	*
Jute, Linens, Cord, and Twine		†	**
Gum and Wood Chemicals		†	**
All Other Industries	—	2,217	40.3
Total	—	$5,500	100.0%

* Less than 0.05%.

** Less than 0.05%; note that these amounts are not included in the total.

† Purchases of less than $500,000; note that these amounts are not included in the total.

SOURCE: Derived from data published by the Bureau of Labor Statistics, Division of Interindustry Economics.

TABLE A-24. RAILROADS SECTOR SALES: 1947

Buyers	*Rank within Sector*	*Gross Sales*	
		Dollars (Millions)	*% of Total*
Consumer Purchases (Households)	1	$2,680	26.9%
Exports (minus Competitive Imports)	2	683	6.9
Government Purchases	3	285	2.9
Gross Private Capital Formation	4	268	2.7
Eating and Drinking Places	5	254	2.5
Grain Mill Products	6	197	2.0
Food Grains and Feed Crops	7	117	1.2
Meat Animals and Products	8	110	1.1
Miscellaneous Food Products	9	97	1.0
Spinning, Weaving, and Dyeing	10	78	0.8
Saw Mills, Planing, and Veneer Mills	11	72	0.7
Inventory Changes	12	70	0.7
Meat Packing and Wholesale Poultry	13	68	0.7
Bakery Products	14	67	0.7
Poultry and Eggs	15	66	0.7
Farm Dairy Products	15	66	0.7
Apparel	17	55	0.5
Alcoholic Beverages	18	49	0.5
Vegetables and Fruits	19	41	0.4
Leather and Leather Goods	20	37	0.4
Tires, Tubes, and Other Rubber Products	20	37	0.4
Canning, Preserving, and Freezing	22	36	0.4
Processed Dairy Products	23	33	0.3
Medical, Dental, and Other Professional Services	23	33	0.3
Paints and Allied Products	25	28	0.3
Soap and Related Products	26	23	0.2
Tobacco Manufactures	27	22	0.2
Vegetable Oils	27	22	0.2
Sugar	29	19	0.2
All Other Agriculture	30	12	0.1
House Furnishings and Other Nonapparel	30	12	0.1
Cotton	32	11	0.1
Special Textile Products	33	7	0.1
Tobacco	34	5	*
Oil-Bearings Crops	34	5	*
Animal Oils	36	4	*
Jute, Linens, Cord, and Twine	36	4	*
Gum and Wood Chemicals	38	3	*
Fishing, Hunting, and Trapping	39	2	*
Canvas Products	40	†	**
All Other Industries	—	4,280	43.0
Total	—	$9,958	100.0%

* Less than 0.05%.

** Less than 0.05%; note that this amount is not included in the total.

† Purchases of less than $500,000; note that this amount is not included in the total.

SOURCE: Derived from data published by the Bureau of Labor Statistics, Division of Interindustry Economics.

TABLE A-25. TRUCKING SECTOR SALES: 1947

Buyers	Rank within Sector	Gross Sales Dollars (Millions)	% of Total
Consumer Purchases (Households)	1	$925	23.5%
Meat Animals and Products	2	185	4.7
Exports (minus Competitive Imports)	3	150	3.8
Food Grains and Feed Crops	4	97	2.5
Grain Mill Products	5	94	2.4
Gross Private Capital Formation	6	93	2.4
Eating and Drinking Places	7	82	2.1
Government Purchases	8	80	2.0
Farm Dairy Products	9	77	1.9
Meat Packing and Wholesale Poultry	10	75	1.9
Saw Mills, Planing, and Veneer Mills	11	63	1.6
Poultry and Eggs	12	47	1.2
Spinning, Weaving, and Dyeing	13	45	1.1
Miscellaneous Food Products	14	37	0.9
Alcoholic Beverages	15	32	0.8
Vegetables and Fruits	16	29	0.7
Processed Dairy Products	17	28	0.7
Inventory Changes	18	27	0.7
Bakery Products	19	26	0.7
Apparel	20	19	0.5
Leather and Leather Goods	21	17	0.4
Tobacco Manufactures	22	16	0.4
Vegetable Oils	23	12	0.3
Canning, Preserving, and Freezing	24	11	0.3
Medical, Dental, and Other Professional Services	24	11	0.3
Sugar	26	5	0.1
All Other Agriculture	27	4	0.1
Soap and Related Products	27	4	0.1
Paints and Allied Products	27	4	0.1
Tires, Tubes, and Other Rubber Products	27	4	0.1
Cotton	31	3	0.1
Animal Oils	31	3	0.1
Special Textile Products	31	3	0.1
House Furnishings and Other Nonapparel	31	3	0.1
Gum and Wood Chemicals	31	3	0.1
Oil-Bearing Crops	36	2	0.1
Fishing, Hunting, and Trapping	36	2	0.1
Tobacco	38	1	*
Jute, Linens, Cord, and Twine		†	**
Canvas Products		†	**
All Other Industries	—	1,613	41.0
Total	—	$3,932	100.0%

* Less than 0.05%.

** Less than 0.05%; note that these amounts are not included in the total.

† Purchases of less than $500,000; note that these amounts are not included in the total.

SOURCE: Derived from data published by the Bureau of Labor Statistics, Division of Interindustry Economics.

TABLE A-26. WATER AND OTHER TRANSPORTATION SECTOR SALES: 1947

Buyers	Rank within Sector	Gross Sales Dollars (Millions)	% of Total
Exports (minus Competitive Imports)	1	$1,691	40.0%
Consumer Purchases (Households)	2	890	21.0
Government Purchases	3	216	5.1
Miscellaneous Food Products	4	35	0.8
Food Grains and Feed Crops	5	23	0.5
Poultry and Eggs	6	20	0.5
Farm Dairy Products	7	11	0.3
Medical, Dental, and Other Professional Services	8	10	0.2
Gross Private Capital Formation	9	9	0.2
Meat Animals and Products	10	8	0.2
Grain Mill Products	10	8	0.2
House Furnishings and Other Nonapparel	12	7	0.2
All Other Agriculture	13	6	0.1
Saw Mills, Planing, and Veneer Mills	13	6	0.1
Vegetables and Fruits	15	5	0.1
Spinning, Weaving, and Dyeing	15	5	0.1
Jute, Linens, Cord, and Twine	15	5	0.1
Eating and Drinking Places	15	5	0.1
Inventory Changes	19	4	0.1
Cotton	20	3	0.1
Meat Processing and Wholesale Poultry	20	3	0.1
Special Textile Products	20	3	0.1
Apparel	20	3	0.1
Paints and Allied Products	20	3	0.1
Oil-Bearing Crops	25	2	*
Processed Dairy Products	25	2	*
Canning, Preserving, and Freezing	25	2	*
Bakery Products	25	2	*
Alcoholic Beverages	25	2	*
Leather and Leather Goods	25	2	*
Soap and Related Products	25	2	*
Tires, Tubes, and Other Rubber Products	25	2	*
Tobacco	33	1	*
Sugar	33	1	*
Tobacco Manufactures	33	1	*
Vegetable Oils	33	1	*
Fishing, Hunting, and Trapping		†	**
Animal Oils		†	**
Canvas Products		†	**
Gum and Wood Chemicals		†	**
All Other Industries	—	1,232	29.1
Total	—	$4,231	100.0%

* Less than 0.05%.

** Less than 0.05%; note that these amounts are not included in the total.

† Purchases of less than $500,000; note that these amounts are not included in the total.

SOURCE: Derived from data published by the Bureau of Labor Statistics, Division of Interindustry Economics.

Table A-27. Warehouse and Storage Sector Sales: 1947

Buyers	Rank within Sector	Gross Sales: Dollars (Millions)	Gross Sales: % of Total
Consumer Purchases (Households)	1	$215	39.9%
Exports (minus Competitive Imports)	2	26	4.8
Meat Packing and Wholesale Poultry	3	23	4.2
Spinning, Weaving, and Dyeing	4	22	4.1
Meat Animals and Products	5	12	2.2
Grain Mill Products	6	9	1.5
Eating and Drinking Places	6	9	1.5
Farm Dairy Products	8	8	1.5
Miscellaneous Food Products	9	6	1.1
Vegetable Oils	10	4	0.7
Poultry and Eggs	11	3	0.6
Food Grains and Feed Crops	11	3	0.6
Bakery Products	11	3	0.6
Leather and Leather Goods	11	3	0.6
Inventory Changes	11	3	0.6
Processed Dairy Products	16	2	0.4
Canning, Preserving, and Freezing	16	2	0.4
Government Purchases	16	2	0.4
Gross Private Capital Formation	16	2	0.4
Vegetables and Fruits	20	1	0.2
Sugar	20	1	0.2
Alcoholic Beverages	20	1	0.2
Apparel	20	1	0.2
House Furnishings and Other Nonapparel	20	1	0.2
Paints and Allied Products	20	1	0.2
Medical, Dental, and Other Professional Services	20	1	0.2
Tires, Tubes, and Other Rubber Products	20	1	0.2
Cotton		†	**
Tobacco		†	**
Oil-Bearing Crops		†	**
All Other Agriculture		†	**
Fishing, Hunting, and Trapping		†	**
Tobacco Manufactures		†	**
Animal Oils		†	**
Special Textile Products		†	**
Jute, Linens, Cord, and Twine		†	**
Canvas Products		†	**
Soap and Related Products		†	**
Gum and Wood Chemicals		†	**
Saw Mills, Planing, and Veneer Mills		†	**
All Other Industries	—	174	32.3
Total	—	$539	100.0%

** Less than 0.05%; note that these amounts are not included in the total.

† Purchases of less than $500,000; note that these amounts are not included in the total.

Source: Derived from data published by the Bureau of Labor Statistics, Division of Interindustry Economics.

TABLE A-28. WHOLESALE TRADE SECTOR SALES: 1947

Buyers	*Rank within Sector*	*Gross Sales* *Dollars (Millions)*	*% of Total*
Consumer Purchases (Households)	1	$6,224	38.6%
Gross Private Capital Formation	2	1,379	8.6
Eating and Drinking Places	3	1,050	6.5
Exports (minus Competitive Imports)	4	1,037	6.4
Apparel	5	309	1.9
Medical, Dental, and Other Professional Services	6	216	1.3
Spinning, Weaving, and Dyeing	7	188	1.2
Food Grains and Feed Crops	8	173	1.1
Meat Packing and Wholesale Poultry	9	119	0.7
Inventory Changes	10	113	0.7
Poultry and Eggs	11	104	0.6
Miscellaneous Food Products	12	92	0.6
Meat Animals and Products	13	76	0.5
Farm Dairy Products	14	70	0.4
Vegetable Oils	14	70	0.4
House Furnishings and Other Nonapparel	16	63	0.4
Vegetables and Fruits	17	62	0.4
Processed Dairy Products	17	62	0.4
Grain Mill Products	19	60	0.4
Leather and Leather Goods	20	59	0.4
Tires, Tubes, and Other Rubber Products	21	58	0.4
Bakery Products	22	55	0.3
Tobacco Manufactures	23	41	0.3
Government Purchases	24	37	0.2
All Other Agriculture	25	33	0.2
Alcoholic Beverages	26	25	0.2
Canning, Preserving, and Freezing	27	24	0.1
Saw Mills, Planing, and Veneer Mills	27	24	0.1
Cotton	29	20	0.1
Special Textile Products	30	18	0.1
Paints and Allied Products	31	14	0.1
Jute, Linens, Cord, and Twine	32	13	0.1
Oil-Bearing Crops	33	12	0.1
Sugar	34	11	0.1
Soap and Related Products	35	9	0.1
Tobacco	36	6	*
Fishing, Hunting, and Trapping	36	6	*
Canvas Products	38	4	*
Animal Oils	39	1	*
Gum and Wood Chemicals	39	1	*
All Other Industries	—	4,164	25.9
Total	—	$16,102	100.0%

* Less than 0.05%.

SOURCE: Derived from data published by the Bureau of Labor Statistics, Division of Interindustry Economics.

TABLE A-29. RETAIL TRADE SECTOR SALES: 1947

Buyers	Rank within Sector	Gross Sales Dollars (Millions)	Gross Sales % of Total
Consumer Purchases (Households)	1	$21,510	83.8%
Gross Private Capital Formation	2	1,049	4.1
Poultry and Eggs	3	234	0.9
Food Grains and Feed Crops	4	187	0.7
Meat Animals and Products	5	141	0.6
Farm Dairy Products	6	125	0.5
Vegetables and Fruits	7	56	0.2
All Other Agriculture	8	42	0.2
Miscellaneous Food Products	9	38	0.1
Medical, Dental, and Other Professional Services	10	28	0.1
Cotton	11	20	0.1
Oil-Bearing Crops	12	17	0.1
Eating and Drinking Places	13	11	*
Tobacco	14	5	*
Fishing, Hunting, and Trapping	15	2	*
Meat Packing and Wholesale Poultry	15	2	*
Exports (minus Competitive Imports)	15	2	*
Grain Mill Products	18	1	*
Bakery Products	18	1	*
Alcoholic Beverages	18	1	*
Apparel	18	1	*
Saw Mills, Planing, and Veneer Mills	18	1	*
Tires, Tubes, and Other Rubber Products	18	1	*
Processed Dairy Products		†	**
Sugar		†	**
Vegetable Oils		†	**
Animal Oils		†	**
Spinning, Weaving, and Dyeing		†	**
Special Textile Products		†	**
Jute, Linens, Cord, and Twine		†	**
Canvas Products		†	**
Leather and Leather Goods		†	**
Soap and Related Products		†	**
Paints and Allied Products		†	**
Gum and Wood Chemicals		†	**
Government Purchases		†	**
All Other Industries	—	2,183	8.5
Total	—	$25,658	100.0%

* Less than 0.05%.

** Less than 0.05%; note that these amounts are not included in the total.

† Purchases of less than $500,000; note that these amounts are not included in the total.

SOURCE: Derived from data published by the Bureau of Labor Statistics, Division of Interindustry Economics.

TABLE A-30. ADVERTISING SECTOR SALES: 1947

Buyers	*Rank within Sector*	*Gross Sales* *Dollars (Millions)*	*% of Total*
Miscellaneous Food Products	1	$182	4.8%
Soap and Related Products	2	122	3.2
Alcoholic Beverages	3	120	3.1
Tobacco Manufactures	4	97	2.5
Apparel	5	85	2.2
Grain Mill Products	6	75	2.0
Canning, Preserving, and Freezing	7	64	1.7
Eating and Drinking Places	8	55	1.4
Consumer Purchases (Households)	9	49	1.3
Spinning, Weaving, and Dyeing	10	34	0.9
Leather and Leather Goods	11	28	0.7
Meat Packing and Wholesale Poultry	12	26	0.7
Tires, Tubes, and Other Rubber Products	13	20	0.5
Bakery Products	14	16	0.4
Processed Dairy Products	15	14	0.4
Government Purchases	16	13	0.3
Paints and Allied Products	17	12	0.3
Medical, Dental, and Other Professional Services	18	11	0.3
All Other Agriculture	19	8	0.2
Saw Mills, Planing, and Veneer Mills	19	8	0.2
Special Textile Products	21	7	0.2
House Furnishings and Other Nonapparel	22	6	0.2
Exports (minus Competitive Imports)	23	3	0.1
Sugar	24	2	0.1
Vegetable Oils	24	2	0.1
Animal Oils	26	1	*
Meat Animals and Products		†	**
Poultry and Eggs		†	**
Jute, Linens, Cord, and Twine		†	**
Canvas Products		†	**
Gum and Wood Chemicals		†	**
All Other Industries	—	2,750	72.2
Total	—	$3,810	100.0%

* Less than 0.05%.

** Less than 0.05%; note that these amounts are not included in the total.

† Purchases of less than $500,000; note that these amounts are not included in the total.

SOURCE: Derived from data published by the Bureau of Labor Statistics, Division of Interindustry Economics.

TABLE A-31. REAL ESTATE AND RENTALS SECTOR SALES: 1947

Buyers	Rank within Sector	Gross Sales Dollars (Millions)	Gross Sales % of Total
Consumer Purchases (Households)	1	$20,289	70.1%
Food Grains and Feed Crops	2	1,437	4.9
Gross Private Capital Formation	3	804	2.8
Eating and Drinking Places	4	386	1.3
Cotton	5	275	0.9
Medical, Dental, and Other Professional Services	6	263	0.9
Government Purchases	7	198	0.7
Meat Animals and Products	8	192	0.7
Oil-Bearing Crops	9	143	0.5
Farm Dairy Products	10	89	0.3
Vegetables and Fruits	11	87	0.3
Apparel	11	87	0.3
Tobacco	13	78	0.3
All Other Agriculture	14	53	0.2
Poultry and Eggs	15	38	0.1
Miscellaneous Food Products	16	31	0.1
Spinning, Weaving, and Dyeing	17	20	0.1
Leather and Leather Goods	18	19	0.1
Processed Dairy Products	19	13	*
Bakery Products	19	13	*
Canning, Preserving, and Freezing	21	11	*
Meat Packing and Wholesale Poultry	22	10	*
House Furnishings and Other Nonapparel	22	10	*
Tires, Tubes, and Other Rubber Products	22	10	*
Saw Mills, Planing, and Veneer Mills	25	8	*
Alcoholic Beverages	26	5	*
Paints and Allied Products	26	5	*
Grain Mill Products	26	5	*
Sugar	29	4	*
Special Textile Products	30	3	*
Tobacco Manufactures	31	2	*
Soap and Related Products	31	2	*
Fishing, Hunting, and Trapping	33	1	*
Vegetable Oils	33	1	*
Animal Oils	33	1	*
Jute, Linens, Cord, and Twine	33	1	*
Canvas Products	33	1	*
Gum and Wood Chemicals	33	1	*
All Other Industries	—	4,337	15.0
Total	—	$28,933	100.0%

* Less than 0.05%.

SOURCE: Derived from data published by the Bureau of Labor Statistics, Division of Interindustry Economics.

Table A-32. Maintenance and Construction Sector Sales: 1947

Buyers	*Rank within Sector*	*Gross Sales* Dollars (Millions)	*% of Total*
Government Purchases	1	$1,936	21.5%
Consumer Purchases (Households)	2	154	1.7
Farm Dairy Products	3	98	1.1
Medical, Dental, and Other Professional Services	4	79	0.9
Eating and Drinking Places	5	73	0.8
Food Grains and Feed Crops	6	40	0.5
Tobacco	7	36	0.4
Spinning, Weaving, and Dyeing	8	23	0.3
Poultry and Eggs	9	20	0.2
Meat Packing and Wholesale Poultry	10	19	0.2
Processed Dairy Products	11	13	0.1
Miscellaneous Food Products	11	13	0.1
Leather and Leather Goods	11	13	0.1
Cotton	14	12	0.1
Apparel	15	11	0.1
Meat Animals and Products	16	10	0.1
Bakery Goods	16	10	0.1
Canning, Preserving, and Freezing	18	9	0.1
Alcoholic Beverages	18	9	0.1
Tires, Tubes, and Other Rubber Products	20	8	0.1
Oil-Bearing Crops	21	7	0.1
All Other Agriculture	21	7	0.1
Grain Mill Products	23	6	0.1
Sugar	24	5	0.1
Saw Mills, Planing, and Veneer Mills	24	5	0.1
Vegetables and Fruits	26	4	*
Special Textile Products	27	3	*
Vegetable Oils	28	2	*
House Furnishings and Other Nonapparel	28	2	*
Soap and Related Products	28	2	*
Paints and Allied Products	28	2	*
Tobacco Manufactures	32	1	*
Animal Oils	32	1	*
Jute, Linens, Cord, and Twine	32	1	*
Canvas Products		†	**
Gum and Wood Chemicals		†	**
All Other Industries	—	6,362	70.7
Total	—	$8,996	100.0%

* Less than 0.05%.

** Less than 0.05%; note that these amounts are not included in the total.

† Purchases of less than $500,000; note that these amounts are not included in the total.

Source: Derived from data published by the Bureau of Labor Statistics, Division of Interindustry Economics.

TABLE A-33. ALL OTHER INDUSTRIES AGGREGATE SALES: 1947

Buyers	*Rank within Aggregate*	*Gross Sales* *Dollars (Millions)*	*% of Total*
Consumer Purchases (Households)	1	$53,639	30.3%
Gross Private Capital Formation	2	29,487	16.6
Government Purchases	3	11,403	6.4
Exports (minus Competitive Imports)	4	4,032	2.3
Inventory Changes	5	1,505	0.8
Saw Mills, Planing, and Veneer Mills	6	864	0.5
Medical, Dental, and Other Professional Services	7	546	0.3
Eating and Drinking Places	8	412	0.2
Apparel	9	297	0.2
Food Grains and Feed Crops	10	209	0.1
Paints and Allied Products	11	186	0.1
Spinning, Weaving, and Dyeing	12	141	0.1
Animal Oils	13	120	0.1
Tires, Tubes, and Other Rubber Products	14	114	0.1
Miscellaneous Food Products	15	111	0.1
Meat Animals and Products	16	98	0.1
Leather and Leather Goods	17	97	0.1
Grain Mill Products	18	89	*
Meat Packing and Wholesale Poultry	19	86	*
Vegetables and Fruits	20	82	*
All Other Agriculture	21	64	*
Canning, Preserving, and Freezing	22	60	*
Farm Dairy Products	23	54	*
Soap and Related Products	24	48	*
Alcoholic Beverages	25	42	*
House Furnishings and Other Nonapparel	25	42	*
Bakery Products	27	37	*
Fishing, Hunting, and Trapping	28	34	*
Cotton	29	30	*
Poultry and Eggs	30	28	*
Processed Dairy Products	30	28	*
Gum and Wood Chemicals	32	20	*
Oil-Bearing Crops	33	19	*
Sugar	34	18	*
Vegetable Oils	34	18	*
Jute, Linens, Cord, and Twine	36	16	*
Special Textile Products	37	15	*
Tobacco	38	12	*
Tobacco Manufactures	39	10	*
Canvas Products	40	3	*
All Other Industries	—	73,118	41.2
Total	—	$177,234	100.0%

* Less than 0.05%.

SOURCE: Derived from data published by the Bureau of Labor Statistics, Division of Interindustry Economics.

TABLE A-34. IMPORTS (NONCOMPETITIVE) SECTOR SALES: 1947

Buyers	*Rank within Sector*	*Gross Sales* Dollars (Millions)	*Gross Sales* % of Total
Government Purchases	1	$2,345	50.9%
Consumer Purchases (Households)	2	833	18.1
Miscellaneous Food Products	3	821	17.8
House Furnishings and Other Nonapparel	4	89	1.9
Jute, Linens, Cord, and Twine	5	50	1.1
Special Textile Products	6	35	0.7
Spinning, Weaving, and Dyeing	7	16	0.3
Food Grains and Feed Crops	8	4	0.1
All Other Agriculture	9	2	*
Apparel	10	1	*
Fishing, Hunting, and Trapping		†	**
All Other Industries	—	432	9.4
Inventory Changes	—	—20	—0.4
Total	—	$4,608	100.0%

* Less than 0.05%.

** Less than 0.05%; note that this amount is not included in the total.

† Purchases of less than $500,000; note that this amount is not included in the total.

SOURCE: Derived from data published by the Bureau of Labor Statistics, Division of Interindustry Economics.

TABLE A-35. GROSS DOMESTIC SALES BY SECTORS: 1947

Sellers	Rank within the National Economy	Gross Sales: Dollars (Millions)	Gross Sales: % of Respective Aggregate†	Gross Sales: % of Total National Economy
Labor and Capital Utilization	1	$228,896		31.2%
Government Services (Taxes, etc.)	2	63,564		8.7
Food Processing Aggregate	3	42,067		5.7
Meat Packing and Wholesale Poultry		11,194	26.6%	1.5
Miscellaneous Food Products		6,631	15.7	0.9
Grain Mill Products		5,342	12.7	0.7
Processed Dairy Products		3,648	8.7	0.5
Bakery Products		3,353	8.0	0.4
Canning, Preserving, and Freezing		2,726	6.5	0.4
Alcoholic Beverages		2,724	6.5	0.4
Tobacco Manufactures		2,567	6.1	0.3
Vegetable Oils		1,921	4.6	0.3
Sugar		1,180	2.8	0.2
Animal Oils		781	1.8	0.1
Farming Aggregate	4	40,274		5.5
Food Grains and Feed Crops		11,003	27.3	1.5
Meat Animals and Products		9,803	24.3	1.3
Farm Dairy Products		5,062	12.6	0.7
Vegetables and Fruits		4,013	10.0	0.6
Poultry and Eggs		3,865	9.6	0.5
Cotton		2,222	5.5	0.3
All Other Agriculture		1,956	4.9	0.3
Oil-Bearing Crops		1,061	2.6	0.1
Tobacco		884	2.2	0.1
Fishing, Hunting, and Trapping		405	1.0	0.1
Real Estate and Rentals	5	28,933		3.9
Fiber Processing Aggregate	6	26,140		3.6
Apparel		11,334	43.4	1.6
Spinning, Weaving, and Dyeing		8,094	30.9	1.1
Leather and Leather Goods		3,733	14.3	0.5
House Furnishings and Other Nonapparel		1,806	6.9	0.2
Special Textile Products		821	3.1	0.1
Jute, Linens, Cord, and Twine		255	1.0	*
Canvas Products		97	0.4	*
Retail Trade	7	25,658		3.5
Wholesale Trade	8	16,102		2.2
Other Chemical Industries	9	10,726		1.5
Container Aggregate	10	10,389		1.4
Paper and Converted Paper Products		6,229	60.0	0.8
Metal Container Materials		1,724	16.6	0.2
Glass		1,153	11.1	0.2
Tin Cans and Other Tins		695	6.7	0.1
Wood Containers and Cooperage		588	5.6	0.1
Railroads	11	9,958		1.4
Maintenance and Construction	12	8,996		1.2

* Less than 0.05%.

† Sectors which are not combined into Aggregates are valued at 100.0%.

TABLE A-35. GROSS DOMESTIC SALES BY SECTORS: 1947 (continued)

Sellers	Rank within the National Economy	Gross Sales: Dollars (Millions)	Gross Sales: % of Total National Economy
Petroleum Products	13	7,572	1.0
Other Fuels (Coal, Coke, and Gas)	14	5,958	0.8
Automobile and Other Repair Services	15	5,500	0.8
Imports (Noncompetitive)	16	4,608	0.6
Electric Light and Power	17	4,438	0.6
Water and Other Transportation	18	4,231	0.6
Trucking	19	3,932	0.5
Advertising	20	3,810	0.5
Tires, Tubes, and Other Rubber Products	21	2,998	0.4
Warehouse and Storage	22	539	0.1
Fertilizers	23	521	0.1
All Other Industries	—	177,234	24.2
Total Gross Domestic Output	—	$733,044	100.0%

SOURCE: Derived from data published by the Bureau of Labor Statistics, Division of Interindustry Economics.

TABLE A-36. PERCENTAGE DISTRIBUTION OF LABOR AND CAPITAL COSTS BY SECTORS OF THE FARMING AGGREGATE:* 1947

Column Number†	Sector	Total Labor and Capital Utilization Inputs‡ (Millions)	% Distribution of Labor and Capital: Wages and Salaries	% Distribution of Labor and Capital: Capital Items§
1	Meat Animals and Products	$3,270	7.4%	92.6%
2	Poultry and Eggs	323	4.9	95.1
3	Farm Dairy Products	1,811	19.1	80.9
4	Food Grains and Feed Crops	6,273	13.6	86.4
5	Cotton	1,501	24.2	75.8
6	Tobacco	656	13.2	86.8
7	Oil-Bearing Crops	621	7.1	92.9
8	Vegetables and Fruits	2,869	29.7	70.3
9	All Other Agriculture	1,386	42.7	57.3
10	Fishing, Hunting, and Trapping	301	‖	‖
11	Farming Aggregate	$19,011	18.4%	81.6%

* The percentages represent a breakdown of the Labor and Capital Utilization inputs as designated in row 58 in Exhibit 2.

† Column numbers refer to same columns as in Exhibit 2.

‡ Obtained from row 58 in Exhibit 2.

§ Capital items include interest, donations, management services, profits, and depreciation.

‖ Data not available.

TABLE A-37. PERCENTAGE DISTRIBUTION OF LABOR AND CAPITAL COSTS BY SECTORS OF THE FOOD PROCESSING AGGREGATE:* 1947

Column Number†	*Sector*	*Total Labor and Capital Utilization Inputs‡ (Millions)*	*% Distribution of Labor and Capital*	
			Wages and Salaries	*Capital Items§*
12	Meat Packing and Wholesale Poultry	$1,579	63.1%	36.9%
13	Processed Dairy Products	615	50.0	50.0
14	Canning, Preserving, and Freezing	687	62.4	37.6
15	Grain Mill Products	796	43.9	56.1
16	Bakery Products	1,248	70.3	29.7
17	Miscellaneous Food Products	1,572	53.2	46.8
18	Sugar	160	56.3	43.7
19	Alcoholic Beverages	1,073	42.7	57.3
20	Tobacco Manufactures	477	50.3	49.7
21	Vegetable Oils	243	33.3	66.7
22	Animal Oils	126	42.1	57.9
23	Food Processing Aggregate	$8,576	44.9%	55.1%

* The percentages represent a breakdown of the Labor and Capital Utilization inputs as designated in row 58 in Exhibit 2.

† Column numbers refer to same columns as in Exhibit 2.

‡ Obtained from row 58 in Exhibit 2.

§ Capital items include interest, donations, management services, profits, and depreciation.

TABLE A-38. PERCENTAGE DISTRIBUTION OF LABOR AND CAPITAL COSTS BY SELECTED SECTORS OF THE FIBER PROCESSING AGGREGATE:* 1947

Column Number†	*Sector*	*Total Labor and Capital Utilization Inputs‡ (Millions)*	*% Distribution of Labor and Capital*	
			Wages and Salaries	*Capital Items§*
24	Spinning, Weaving, and Dyeing	$3,025	63.9%	36.1%
25	Special Textile Products	356	53.3	46.7
	Total of Above Two Sectors	$3,381	63.7%	36.3%

* This table represents only two industry sectors of the Fiber Processing Aggregate and cannot be illustrative of the whole aggregate. The sum of the above sector inputs from Labor and Capital Utilization (row 58) represents approximately one-third of all the inputs from Labor and Capital Utilization to the Fiber Processing Aggregate. The percentages represent a breakdown of the Labor and Capital Utilization inputs as designated in row 58, Exhibit 2.

† Column numbers refer to same columns as in Exhibit 2.

‡ Obtained from row 58 in Exhibit 2.

§ Capital items include interest, donations, management services, profits, and depreciation.

APPENDIX III

SECTION A

DESCRIPTION OF CONSUMER PURCHASES COLUMN 44 FOR AGRIBUSINESS INPUT-OUTPUT CHART (EXHIBIT 2)

The most important final demand sector is that of Consumer Purchases (column 44). For the sake of clarity let us examine column 44 and trace selected input-output relationships pertaining to it from top to bottom. Consumers purchased directly from each of the sectors at the left of the chart, without any intermediate processing, the following items: $1,070 million from Meat Animals and Products (row 1) [most of this was on-the-farm consumption, primarily hog consumption]; $2,589 million from Poultry and Eggs (row 2) [most of this was egg consumption that required no processing]; $2,731 million Farm Dairy Products (row 3) [this was on-the-farm consumption and general consumer consumption of fluid milk products]; $45 million from the Food Grains and Feed Crops sector (row 4) [mostly in the form of sweet corn]; no direct purchases from the Cotton or Tobacco sectors (rows 5 and 6) [because the total output required processing]; $3 million from the Oil-Bearing Crops sector (row 7) [primarily nuts]; $2,651 million from the Vegetables and Fruits sector (row 8) [fresh vegetables and fruits]; $641 million from the All Other Agriculture sector (row 9); and $85 million directly from the Fishing, Hunting, and Trapping sector (row 10). This makes a total of $9,815 million of end products purchased directly by consumers from the sectors comprising the Farming Aggregate (row 11).

Continuing the analysis of column 44, one notes that consumers purchased $23,728 million from sectors contained in the Food Processing Aggregate (row 23); $13,425 million from those of the Fiber Processing Aggregate (row 31); $725 million from those of the Container Aggregate (row 37); $8 million purchased from the Fertilizer sector (row 38); and $2,576 million from the Power Aggregate (row 42) [mostly in the form of petroleum products]. The remaining sectors from whom consumers purchased items directly include the following: the Other Chemical Industries sector, $1,988 million (row 43); Tires, Tubes, and Other Rubber Products, $731 million (row 44); Automobile and Other Repair Services, $2,585 million (row 45); Railroads, $2,680 million (row 46) [including amounts spent on railroad fares and railroad transportation costs of the goods directly purchased by consumers]; Trucking, $925 million (row 47) [including trucking costs of goods directly purchased by consumers]; Water and Other Transportation, $890 million (row 48) [including amounts spent on air and water travel for passenger fares and transportation costs of the goods directly purchased by consumers]; Warehouse and Storage, $215 million (row 49) [including warehousing and storage for personal use by consumers]; Wholesale Trade, $6,224 million (row 50) [including wholesale trade margins on all consumer purchases stated in producers' values]; Retail Trade, $21,510 million (row 51) [including all retail trade margins on all consumer purchases stated in producers' values]; Advertising, $49 million (row 52) [including advertising services purchased by individuals to popularize products or services]; Real Estate and Rentals, $20,289 million (row 53) [including all space rentals of all tenant-occupied farm and nonfarm dwellings]; and Maintenance and Construction, $154 million (row 54) [including maintenance and repair of old construction for consumer purchases stated in terms of tenant repairs]. Approximately $16,000 million of *private* construction is included as private capital formation in the final demand sector whereas new *public* construction appears as a government purchase. The sum of $53,639 million (row 55) of the All Other Industries Aggregate is included in the form of purchases by consumers from those processing or intermediate sectors of the economy not previously mentioned.

The remaining three sectors in column 44 comprise factor payment sectors as follows: noncompetitive imports $833 million (row 56) [items which are purchased directly by individuals and which are not produced in this country]; Government Services $32,432 million (row 57) [this in essence consists of personal income taxes paid by individuals for government operations or services which include national defense, welfare, agriculture, commerce, labor, etc.]; Labor and Capital Utilization $2,116 million (row 58) [this includes wages and salaries—in the case of consumer purchases paid primarily to household help—interest payments, entrepreneurial income, royalties,[1] and contributions to private pension plans]. The final sector row in column 44 is a summation of outlays by consumers in 1947 to processing and factor payments sectors, totaling $197,537 million (row 59). If one subtracts taxes paid by consumers in the amount of $32,432 million (row 57) from the grand total of all consumer expenditures of $197,537 million (row 59), one obtains a remainder of $165,105 million, which approximates the total for personal consumption expenditures in purchasers' values (including both wholesale and retail margins) obtained by the Commerce Department staff.

[1] In the case of corporations it also includes corporate profits after taxes.

SECTION B

RECONCILIATION OF THE AGRIBUSINESS FLOW CHART (EXHIBIT 3) WITH THE AGRIBUSINESS INPUT-OUTPUT CHART (EXHIBIT 2)[1]

Exhibit 3—Flow Chart	*Exhibit 2—Source*	
CONSUMER PURCHASES	CONSUMER PURCHASES	
1. Soap and Related Products plus Paints and Allied Products $0.95 billion ($0.92 soaps and $0.03 paints)	1. Part of "All Other Industries Aggregate" (column 44, row 55). These and other sectors were separated out only as purchasing sectors or columns because their main relationships within agribusiness are as markets for agricultural products and not as important suppliers to the Farming Aggregate	
2. Leather and Leather Products $2.07 billion	2. Column 44, row 30.	
		In millions
3. Food Industries $21.02 billion	3. Column 44, row 23	= $23,728
	minus (Alcoholic Beverages, column 44, row 19; Tobacco Manufactures, column 44, row 20; Vegetable Oils, column 44, row 21; Animal Oils, column 44, row 22)	= 2,713
		$21,015
4. Nonprocessed Foods $9.73 billion	4. Column 44, row 11	= $ 9,815
	minus Fishing, Hunting and Trapping, column 44, row 10	= 85
		$ 9,730
5. Drinking and Eating Places $13.11 billion	5. See 1 above.	
6. Alcoholic Beverages $1.21 billion	6. Column 44, row 19	= $ 1,209
7. Textiles $11.36 billion	7. Column 44, row 31	= $13,425
	minus Leather and Leather Goods, column 44, row 30	= 2,065
		$11,360
8. Tobacco Products $1.48 billion	8. Column 44, row 20	= $ 1,485
9. Wood and Paper Products $1.99 billion	9. See 1 above.	
10. Wholesale and Retail Trade Margins $8.5 billion; All Other $1.5 billion	10. Estimate based on part of the $6,224 million total consumer wholesale trade margins (column 44, row 50) and $21,510 retail trade margins (column 44, row 51)	
11. Total $72.92 billion	11. Total $72.92 billion	

[1] All figures in producers' values unless otherwise stated.

INTERMEDIATE ACTIVITIES

12. Soap and Paint purchased $0.72 billion from oil processors

13. Leather and Leather Products purchased $0.49 billion from Food Industries

14. Vegetable and Animal Oils purchased $0.81 billion from Agriculture; sold $0.93 billion to Food Industries, $0.21 billion of oil meals to Agriculture

15. Food Industries purchased $14.54 billion from Agriculture and sold $2.19 billion to Agriculture mostly in the form of bran and midds. Food Industries also sold $2.55 billion to Drinking and Eating Places

16. Drinking and Eating Places purchased $0.79 billion from Agriculture

17. Alcoholic Beverages purchased $0.35 billion from Agriculture
    Alcoholic Beverages sold $0.99 billion to Drinking and Eating Places

18. Textiles purchased $2.09 billion from Agriculture

19. Tobacco Products purchased $0.78 billion from Agriculture

20. Wood and Paper Processing purchased $0.19 billion from Agriculture

INTERMEDIATE ACTIVITIES

12. Soap purchased from vegetable oils (column 32, row 21)	=	$ 150
and purchased from animal oils (column 32, row 22)	=	374
Paints purchased from vegetable oils (column 33, row 21)	=	188
and purchased from animal oils (column 33, row 22)	=	8
		$ 720
13. Leather and Leather Goods purchased from Meat Animals and Products (column 30, row 1)	=	$ 47
and from Meat Packing and Wholesale Poultry (column 30, row 12)	=	443
		$ 490
14. Vegetable and Animal Oils purchased from Oil-Bearing Crops (column 21, row 7)	=	$ 815
sold to Food Processing (column 23, row 7)	=	934
sold to Farming Aggregate (column 11, rows 21 and 22)	=	209
15. Food Processing purchased from Farming (column 23, row 11)	=	$16,913
minus (Alcoholic Beverages, column 19, row 11; Tobacco Manufactures, column 20, row 11; Vegetable Oils, column 21, row 11; Hunting, Fishing, and Trapping, column 23, row 10)	=	2,374
		$14,539
Food Processing sold to Grain Mill Products (column 11, row 15)	=	$ 2,263
minus Food Grains and Feed Crops purchases (column 4, row 15)	=	77
		$ 2,186
Food Processing sold to Drinking and Eating Places (column 36, row 23)	=	$ 3,535
minus Alcoholic Beverages (column 36, row 19)	=	986
		$ 2,549
16. Column 36, row 11	=	$ 865
minus column 36, row 10	=	73
		$ 792
17. Column 19, row 11	=	$ 348
Column 36, row 19	=	$ 986
18. Column 31, row 11	=	$ 2,382
minus column 31, row 10	=	241
column 30, row 11	=	51
		$ 2,090
19. Column 20, row 11	=	$ 783
20. Column 35, row 11	=	$ 188

AGRICULTURE PURCHASES	AGRICULTURE PURCHASES	
21. Agriculture purchased $2.42 billion from Feed Manufactures	21. Estimates from U.S.D.A. and column 11, row 23	= $ 2,543
22. Agriculture purchased $0.5 billion from Seed Supplies	22. Estimates from U.S.D.A.	
23. Agriculture purchased, repaired, and depreciated $3.6 billion of farm machinery and automotive power	23. This amount was not estimated in the input-output data used in Exhibit 2, as it is included in the Labor and Capital Utilization sector as an aggregate item of $19,011 million (see column 11, row 58) as well as part of Gross Private Capital Formation for the farm machinery industry sector included in All Other Industries (see column 42, row 55)	
24. The wholesale margin on the items purchased by Agriculture was $0.6 billion	24. Column 11, row 50 minus column 10, row 50	= $ 562 = 6 $ 556
25. The retail margin on the items purchased by Agriculture was $0.8 billion	25. Column 11, row 51 minus column 10, row 51	= $ 829 = 2 $ 827
26. Agriculture purchased $2.08 billion from all other sources	26. Column 11, row 55 plus U.S.D.A. estimates	= $ 630 = 1,450 $ 2,080
27. Agriculture purchased $0.96 billion from Transportation	27. Column 11, rows 46, 47, 48 minus column 10, rows 46, 47, 48	= $ 961 = 4 $ 957
28. Agriculture purchased $0.5 billion from Power	28. Column 11, row 42	= $ 521
29. Agriculture purchased $1 billion from Containers	29. Column 11, row 37 plus U.S.D.A. estimates	= $ 169 = 831 $ 1,000
30. Agriculture purchased $0.42 billion from Fertilizers	30. Column 11, row 38	= $ 415

SECTION C

AGRIBUSINESS CAPITAL, LABOR, AND INVENTORY COEFFICIENTS

Closely related to the flow of goods and services through the economy are such items as capital, labor, and inventories. The Harvard Economic Research Project, under the direction of Dr. Wassily Leontief and Dr. Elizabeth Gilboy, and with the help of specialists in various fields of study such as Dr. James M. Henderson—capital coefficients, Dr. Leon N. Moses—interregional analysis, Charlotte Taskier—inventory coefficients, and Alfred H. Conrad—labor coefficients, has analyzed resources and labor in a manner which supplements the 1947 input-output study.

Table A–39, capital coefficients for agribusiness, adapted from the Harvard Study, indicates the amount of capital required to produce a given gross output, assuming a continuation of the pattern of technology and production that existed in 1947. The term coefficient, as used here, is defined as that dollar value of the products of a given capital-supplying industry required to produce one dollar of productive capacity. Summarizing Table A–39, we have the following:

Capital Investment per Dollar of Gross Output for Selected Agribusiness Aggregates, 1947

Suppliers	$ .5132
Farming Aggregate	1.6114
Processing and Distributing Industries	.2077

From the above tabulation, one may note that in 1947 every dollar of annual output required $1.61 of capital for the Farming Aggregate, $.51 for the Farm Supplies Aggregate, and $.21 for the Processing-Distribution Aggregate, respectively. Even after allowing for the differences in evaluating capital assets as between Farming and the other aggregates, explained earlier, these data confirm the high fixed-capital investment in the Farming Aggregate noted in the balance sheet analysis, Chapter 2, and also in Table A–40.

In a like manner, the labor required in 1947 by agribusiness industries in producing output to satisfy $1 million of consumer demand is set forth in Table A–41. For example, the Meat Animals sector required directly and indirectly 133.0 and 39.4 man-years, respectively, per $1 million of meat products delivered to consumer demand. The largest requirements of total labor per $1 million output were in the textile industries (including Spinning, Weaving, and Dyeing; Apparel; and Canvas Products) and Poultry and Eggs industry—all of which required in excess of 200 man-years. The lowest use of labor was in Leather Tanning with 79 man-years. Also, there was wide variation between sectors as to the relative importance of direct and indirect labor—ranging from a low of 21% indirect in the case of Jute, Linens, Cord, and Twine to 54% for Farm and Industrial Tractors.

Another phase of agribusiness worthy of note is that of the relative inventory position of its various component parts.[1] Table A–42 indicates that in 1947 six of the ten industry sectors and aggregates in the economy having the largest inventories of raw material and finished goods were within agribusiness. This confirms the fact that agribusiness has relatively large inventories as well as capital requirements. The distribution of inventories among the selected sectors of the economy is presented in Table A–42.[2] With respect to business entities, the term inventories is used in an accounting sense. In the case of the Farming Aggregate, comparable figures are not available. Farm inventories, as presented in Table A–42, are based on average 1946–1950 quantities and values, and include commodities in storage, crops, and livestock less than 3 years old. Admittedly, this base results in a rough figure. However, even after allowing for a wide margin of error, farm inventories loom large compared with those of other sectors of agribusiness or the economy. In view of accumulating surpluses, the use of such figures doubtless would understate the current situation.

[1] Of course, agribusiness inventories are subject to wide variation year by year. No one year can be considered as representative. Also, many details are obscured in the process of aggregating.

[2] More recent input-output studies relating to agriculture have been made by Fox and Norcross, and Peterson and Heady (see footnote 2, p. 25, Chapter 3). Fox and Norcross indicate some of the analyses that can be made by studying the expenditure items in a farming aggregate input-output chart, and the sources of cash farm income for individual farm commodities or groups of commodities. They recognize the limitations of aggregating certain crops, livestock, and agricultural regions, as well as the importance of supplementing quantitative input-output facts with other information concerning supply and demand curves of agricultural commodities. Peterson and Heady deal more extensively with the Farming Aggregate's relations to the total economy by indicating the changing coefficient relations (in that different ratios and relationships exist between the Farming Aggregate and the remainder of the economy) from 1939 to 1949. Some of their conclusions are comparable with those already developed in our agribusiness flow diagrams for 1947 and 1954, namely (1) that there has been an increased dependency of crop-farming enterprises on industry, the coefficients of interdependence having increased from $.36 to $.56 during the period 1939 to 1949, and (2) that a $1.00 increase in demand for "primary" or "crop production" enterprises would require a $.06 increase in "secondary" or livestock enterprises, and a $.42 increase in the output of industry. The same $1.00 increase in "demand" for industrial products would increase output in "crop" agriculture by $.10 and in livestock agriculture by $.08. From this analysis, Peterson and Heady conclude that industry depends little on agriculture, but that agriculture depends heavily on industry for increases in output. However, the farm supplies manufacturers are more directly affected by changes in farmer purchasing power, because farmers are the direct buyers of their products. The direct and indirect impact of a change in demand has been emphasized throughout this study. It is not the size of the particular industry sector or aggregate, but rather its relationship to other parts of the economy, that determines its weight in the nation's progress.

TABLE A-39. CAPITAL COEFFICIENTS BY SELECTED SECTORS OF AGRIBUSINESS: 1947

Sector	*Estimated Capital Stocks* (Millions)*	*Gross Output (Millions)*	*Capital Coefficients†*
Processing and Distributing Industries:			
Meat Packing and Wholesale Poultry	$1,061	$11,106	.0955
Processed Dairy Products	777	3,647	.2131
Canning, Preserving, and Freezing	840	2,725	.3084
Grain Mill Products	626	5,344	.1172
Bakery Products	1,088	3,352	.3244
Miscellaneous Food Products	2,260	6,633	.3407
Sugar	716	1,180	.6063
Alcoholic Beverages	1,014	2,725	.3720
Tobacco Manufactures	272	2,565	.1060
Spinning, Weaving, and Dyeing	2,520	8,096	.3113
Special Textile Products	281	822	.3419
Jute, Linens, Cord, and Twine	87	255	.3411
Canvas Products	12	97	.1204
Apparel	1,301	11,335	.1148
House Furnishings and Other Nonapparel	287	1,806	.1590
Vegetable Oils	264	1,913	.1382
Animal Oils	106	776	.1373
Leather Tanning and Finishing	146	1,075	.1359
Other Leather Products	46	537	.0854
Total Processing and Distributing Industries	$13,704	$65,989	.2077
Suppliers:			
Fertilizers	$262	$523	.5011
Farm and Industrial Tractors	593	1,143	.5188
Total Suppliers	$855	$1,666	.5132
Farming Aggregate (excluding Fishing, Hunting, and Trapping)	$64,241	$39,866	1.6114

* These stocks represent the amount of capital that would have been required to produce the 1947 gross outputs given the technology contained in the capital coefficients.

† A capital coefficient is defined as capital stock divided by gross output, or the dollar value of the products of a capital-supplying industry required to produce one dollar of productive capacity.

SOURCE: *Estimates of the Capital Structure of American Industries, 1947*, Harvard Economic Research Project (June 1953).

TABLE A-40. TWENTY LEADING SECTORS IN TERMS OF CAPITAL STOCKS:* 1947

Sector†	*Rank*	*Estimated Dollars (Millions)*	*Capital Stocks % of Grand Total*
Real Estate and Rentals	1	$236,051	41.51%
Farming Aggregate (excluding Fishing, Hunting, and Trapping)	2	64,241	11.30
Railroads	3	33,254	5.85
Retail Trade	4	31,729	5.58
Nonprofit Institutions	5	26,661	4.69
Electric Light and Power	6	14,631	2.57
Telephone and Telegraph	7	11,942	2.10
Wholesale Trade	8	9,403	1.65
Medical, Dental, and Other Professional Services	9	9,110	1.60
Eating and Drinking Places	10	8,675	1.53
Steel Works and Rolling Mills	11	8,282	1.46
Petroleum Products	12	7,119	1.25
Crude Petroleum and Natural Gas	13	6,654	1.17
Motor Vehicles	14	4,971	0.87
Construction	15	4,010	0.71
Overseas Transportation	16	3,911	0.69
Other Water Transportation	17	3,705	0.65
Motion Pictures and Other Amusements	18	3,161	0.56
Coke and Products	19	2,645	0.47
Printing and Publishing	20	2,582	0.45
Total of Above Sectors	—	$492,736	86.66%
Grand Total of 192 Sectors	—	$568,609	100.00%

* These stocks represent the amount of capital that would have been required to produce the 1947 gross outputs given the technology contained in the capital coefficients.

† Sectors in this table are taken from the more detailed 192 sector breakdown of the Bureau of Labor Statistics.

SOURCE: *Estimates of the Capital Structure of American Industries, 1947*, Harvard Economic Research Project (June 1953).

TABLE A-41. LABOR REQUIREMENTS PER MILLION DOLLARS OF FINAL DEMAND FOR THE OUTPUT OF SELECTED SECTORS: 1947
(Requirements in man-years)

Sector	*Total*	*Direct*	*Indirect*	*Principal Source of Indirect Requirements*
Meat Animals and Products	172.4	133.0	39.4	Food Grains and Feed Crops
Poultry and Eggs	213.1	148.6	64.5	Food Grains and Feed Crops; Grain Mill Products; Real Estate and Rentals
Farm Dairy Products	163.8	121.2	42.6	Food Grains and Feed Crops; Grain Mill Products
Food Grains and Feed Crops	128.6	92.1	36.5	Real Estate and Rentals
Cotton	135.5	102.4	33.1	Real Estate and Rentals
Tobacco	105.8	82.5	23.3	Various
Oil-Bearing Crops	125.9	93.1	32.8	Real Estate and Rentals
Vegetables and Fruits	125.6	85.1	40.5	Various
All Other Agricultural Products	120.5	85.4	35.1	Various
Meat Packing and Wholesale Poultry	178.3	129.9	48.4	Meat Animals and Products; Food Grains and Feed Crops
Processed Dairy Products	173.2	115.3	57.9	Farm Dairy Products; Food Grains and Feed Crops
Canning, Preserving, and Freezing	200.3	106.4	93.9	Vegetables and Fruits
Grain Mill Products	129.4	68.9	60.5	Food Grains and Feed Crops
Bakery Products	187.9	122.2	65.7	Food Grains and Feed Crops; Grain Mill Products; Miscellaneous Food Products
Miscellaneous Food Products	144.5	73.3	71.2	Oil-Bearing Crops; Sugar; Vegetable Oils
Sugar	157.9	108.6	49.3	All Other Agricultural Products
Alcoholic Beverages	125.6	64.8	60.8	Food Grains and Feed Crops
Tobacco Manufactures	146.5	99.8	46.7	Tobacco
Spinning, Weaving, and Dyeing	207.0	152.4	54.6	Cotton
Special Textile Products	182.6	142.3	40.3	Meat Animals and Products; Spinning, Weaving, and Dyeing
Jute, Linens, Cord, and Twine	164.7	129.9	34.8	Cotton; Spinning, Weaving, and Dyeing
Canvas Products	228.0	179.4	48.6	Cotton; Spinning, Weaving, and Dyeing
Apparel	235.9	180.0	55.9	Spinning, Weaving, and Dyeing
House Furnishings and Other Nonapparel	191.4	133.6	57.8	Cotton; Spinning, Weaving, and Dyeing
Fertilizers	174.6	113.4	61.2	Industrial Inorganic Chemicals; Railroads
Vegetable Oils	153.6	95.7	57.9	Oil-Bearing Crops; Miscellaneous Food Products; Real Estate and Rentals
Animal Oils	179.0	100.2	78.8	Meat Animals and Products; Food Grains and Feed Crops; Meat Packing and Wholesale Poultry
Leather Tanning	79.2	54.4	24.8	Various
Other Leather Products	209.6	158.7	50.9	Leather Tanning
Footwear except Rubber	215.4	168.8	46.6	Leather Tanning
Tin Cans and Tin Ware	189.1	119.0	70.1	Blast Furnaces; Steel Works and Rolling Mills
Farm and Industrial Tractors	199.4	92.6	106.8	Steel Works and Rolling Mills
Farm Equipment	208.9	119.3	89.6	Steel Works and Rolling Mills

SOURCE: *Report on Research for 1954*, Harvard Economic Research Project (February 1955).

TABLE A-42. TWENTY LEADING SECTORS IN TERMS OF TOTAL INVENTORIES: 1947

		Raw Materials		*Finished Goods*		*Total Inventories*	
Sector†	*Rank*	*Dollars (Millions)*	*% of Grand Total*	*Dollars (Millions)*	*% of Grand Total*	*Dollars (Millions)*	*% of Grand Total*
Farming Aggregate (excluding Fishing, Hunting, and Trapping)*	1	$ 0	0.00%	$18,521	56.67%	$18,521	34.24%
Motor Vehicles	2	901	4.21	779	2.39	1,680	3.11
Apparel	3	860	4.01	776	2.37	1,636	3.02
Spinning, Weaving, and Dyeing	4	956	4.46	568	1.74	1,524	2.82
Alcoholic Beverages	5	398	1.86	877	2.68	1,275	2.36
Petroleum Products	6	310	1.45	957	2.93	1,267	2.34
Tobacco Manufactures	7	647	3.02	535	1.64	1,182	2.18
Miscellaneous Food Products	8	460	2.15	393	1.20	853	1.57
Railroads	9	795	3.71	0	0.00	795	1.47
Steel Works and Rolling Mills	10	571	2.67	197	0.60	768	1.42
Canning, Preserving, and Freezing	11	216	1.01	540	1.65	756	1.40
Meat Packing and Wholesale Poultry	12	233	1.09	446	1.36	679	1.26
Aircraft and Parts	13	585	2.73	54	0.17	639	1.18
Grain Mill Products	14	488	2.28	134	0.41	622	1.15
Construction, Mining, Oil Field Machinery, and Tools	15	234	1.09	291	0.89	525	0.97
Logging, Saw Mills, Planing, and Veneer Mills	16	166	0.77	349	1.07	515	0.95
Eating and Drinking Places	17	505	2.36	0	0.00	505	0.93
Drugs and Medicines	18	177	0.83	309	0.94	486	0.90
Converted Paper Products	19	213	1.00	263	0.80	476	0.88
Hardware	20	83	0.39	361	1.11	444	0.82
Total of Above 20 Sectors	—	$8,798	41.09%	$26,350	80.62%	$35,148	64.97%
Grand Total of 192 Sectors	—	$21,415	100.00%	$32,683	100.00%	$54,097	100.00%

* Production supplies in farmers' possession have been considered as capital assets. The $18 billion Farming inventory of finished goods is based on the average inventories for Farming from 1946–1950. Also included is the value of crops and all livestock less than 3 years old—older animals being considered as capital assets.

† Sectors in this table are taken from the more detailed 192 sector breakdown of the Bureau of Labor Statistics.

SOURCE: *Estimates of the Capital Structure of American Industries, 1947,* Harvard Economic Research Project (June 1953).

SECTION D

Bakery Products Capital, Labor, and Inventory Coefficients

In 1947 the Bakery Products sector had the following position with respect to capital, labor, and inventories:[1]

Capital Coefficient[2]
$.32 of investment per dollar of output
Labor Requirement[3]
122.2 direct man-years } per million dollars
65.7 indirect man-years } of final demand
Inventory Coefficients
$.0563 raw material per dollar of output[4]
$.0060 finished goods per dollar of output[5]

In 1947 the relative capital position of the baking industry was lower per dollar of output than that of either the Farm Supplies or Farming Aggregates, but more than the average requirements for most of the other Food Processing sectors (see Tables A–39 and A–43).

[1] The labor and inventory data rather accurately indicate the requirements for production, but the capital coefficient is subject to inaccuracy, depending on depreciation policy.
[2] Derived from Tables A-39 and A-43.
[3] Derived from Table A-41.
[4] Derived from Table A-44.
[5] See source to Table A-44.

Table A-43. Components of the Capital Coefficient for Bakery Products:* 1947

Capital-Supplying Industries	*Components of the Coefficient*
Wood Furniture	$.0016
Metal Furniture	.0033
Special Industrial Food Machinery	.0252
Elevators and Conveyors	.0355
Blowers and Fans	.0202
Industrial Machines (Furnaces, etc.)	.0482
Commercial Machines (Computers, etc.)	.0057
Motor Vehicles	.0288
Truck Trailers	.0008
Railroads	.0016
Trucking	.0006
Water Transportation	.0002
New and Maintenance Construction	.1527
Bakery Products Capital Coefficient	$.3244

* Capital coefficients are defined as the dollar value of the stocks of the products of each capital-supplying industry required to produce one dollar of productive capacity.

Source: *Estimates of the Capital Structure of American Industries, 1947,* Harvard Economic Research Project (June 1953).

The labor requirements indicate that the output of Bakery Products in 1947 generated half as many man-hours of employment in related industries as within itself—particularly in such sectors as Food Grains and Feed Crops, Grain Mill Products, and Miscellaneous Food Products. The inventory requirements were quite low, especially the finished goods inventories, reflecting the importance of "freshness" to this industry sector.

Table A-44. Components of the Raw Materials Inventory Coefficient for Bakery Products:* 1947

Raw Materials Supplying Industries	*Components of the Coefficient*
Poultry and Eggs	$.0005
Dairy Farm Products	.0006
Oil-Bearing Crops	.0001
Vegetables and Fruits	.0004
All Other Agriculture (Nuts, etc.)	.0002
Coal Mining	.0001
Meat Packing and Wholesale Poultry ("Over-all")	.0033
Processed Dairy Products	.0028
Canning, Preserving, and Freezing	.0015
Grain Mill Products	.0211
Bakery Products	.0005
Miscellaneous Food Products (Shortening, etc.)	.0112
Sugar	.0033
Converted Paper Products	.0040
Plastics Materials	.0004
Vegetable Oils	.0006
Miscellaneous Chemical Industries	.0002
Petroleum Products	.0005
Tires and Inner Tubes	.0001
Miscellaneous Rubber Products	.0001
Special Industrial Machinery	.0001
Railroads	.0020
Trucking	.0008
Storage	.0001
Water Transportation	.0001
Wholesale Trade	.0017
Bakery Products Raw Materials Coefficient	$.0563

* Raw materials inventory coefficients are defined as the dollar value of the stocks of the products of each raw materials-supplying industry required to produce one dollar of output for every industry.

Source: *Estimates of the Capital Structure of American Industries, 1947,* Harvard Economic Research Project (June 1953).

Printed in the USA
CPSIA information can be obtained
at www.ICGtesting.com
LVHW081140160524
780414LV00022B/370